AF148393

Research Teams and Partnerships: Trends in the Chemical Sciences

REPORT OF A WORKSHOP

Chemical Sciences Roundtable
Board on Chemical Sciences and Technology
Commission on Physical Sciences, Mathematics, and Applications
National Research Council

NATIONAL ACADEMY PRESS
Washington, D.C.

NOTICE: The project that is the subject of this report was approved by the Governing Board of the National Research Council, whose members are drawn from the councils of the National Academy of Sciences, the National Academy of Engineering, and the Institute of Medicine. The members of the workshop organizing committee responsible for the report were chosen for their special competences and with regard for appropriate balance.

Support for this project was provided by National Science Foundation under Grant No. CHE-9630106, the National Institutes of Health under Contract No. N01-OD-4-2139, and the U.S. Department of Energy under Grant No. DE-FG02-95ER14556. Any opinions, findings, conclusions, or recommendations expressed in this material are those of the authors and do not necessarily reflect the views of the National Science Foundation, the National Institutes of Health, or the U.S. Department of Energy.

International Standard Book Number 0-309-06827-4

Additional copies of this report are available from:

National Academy Press
2101 Constitution Avenue, NW
Box 285
Washington, DC 20055
800-624-6242
202-334-3313 (in the Washington metropolitan area)
http://www.nap.edu

Board on Chemical Sciences and Technology
2101 Constitution Avenue, NW
NAS 273
Washington, DC 20418
202-334-2156

THE NATIONAL ACADEMIES

National Academy of Sciences
National Academy of Engineering
Institute of Medicine
National Research Council

The **National Academy of Sciences** is a private, nonprofit, self-perpetuating society of distinguished scholars engaged in scientific and engineering research, dedicated to the furtherance of science and technology and to their use for the general welfare. Upon the authority of the charter granted to it by the Congress in 1863, the Academy has a mandate that requires it to advise the federal government on scientific and technical matters. Dr. Bruce M. Alberts is president of the National Academy of Sciences.

The **National Academy of Engineering** was established in 1964, under the charter of the National Academy of Sciences, as a parallel organization of outstanding engineers. It is autonomous in its administration and in the selection of its members, sharing with the National Academy of Sciences the responsibility for advising the federal government. The National Academy of Engineering also sponsors engineering programs aimed at meeting national needs, encourages education and research, and recognizes the superior achievements of engineers. Dr. William A. Wulf is president of the National Academy of Engineering.

The **Institute of Medicine** was established in 1970 by the National Academy of Sciences to secure the services of eminent members of appropriate professions in the examination of policy matters pertaining to the health of the public. The Institute acts under the responsibility given to the National Academy of Sciences by its congressional charter to be an adviser to the federal government and, upon its own initiative, to identify issues of medical care, research, and education. Dr. Kenneth I. Shine is president of the Institute of Medicine.

The **National Research Council** was organized by the National Academy of Sciences in 1916 to associate the broad community of science and technology with the Academy's purposes of furthering knowledge and advising the federal government. Functioning in accordance with general policies determined by the Academy, the Council has become the principal operating agency of both the National Academy of Sciences and the National Academy of Engineering in providing services to the government, the public, and the scientific and engineering communities. The Council is administered jointly by both Academies and the Institute of Medicine. Dr. Bruce M. Alberts and Dr. William A. Wulf are chairman and vice chairman, respectively, of the National Research Council.

CHEMICAL SCIENCES ROUNDTABLE

COMMISSION ON PHYSICAL SCIENCES, MATHEMATICS, AND APPLICATIONS

Preface

The Chemical Sciences Roundtable (CSR) was established in 1997 by the National Research Council (NRC). It provides a science-oriented, apolitical forum for leaders in the chemical sciences to discuss chemically related issues affecting government, industry, and universities. Organized by the NRC's Board on Chemical Sciences and Technology, the CSR aims to strengthen the chemical sciences by fostering communication among the people and organizations—spanning industry, government, universities, and professional associations—involved with the chemical enterprise. The CSR does this primarily by organizing workshops that address issues in chemical science and technology that require national attention.

At its December 1997 and May 1998 meetings, the CSR identified the topic of research teams and partnerships as an area of broad interest in the chemical sciences community. To provide a forum for exploring this topic, an organizing committee was formed, and a workshop was planned for May 1999.

The aim of the workshop was to discuss the circumstances under which research and development (R&D) teams and partnerships in the chemical sector are appropriate and successful. Issues put forward for discussion included examination of the goals of R&D teams and partnerships, the barriers to implementing them, their potential benefits, and the available methodologies for carrying them out. What are the key trends, how are collaborations changing as a result, and how are best practices evolving? The workshop, "Research Teams and Partnerships: Trends in the Chemical Sciences," brought together research scientists, managers, and technology transfer specialists from government, industry, and academia to discuss and explore the trends and potential impacts of partnerships and teams for the conduct of research in the chemical sciences.

The papers in this volume are the authors' own versions of their presentations, and the discussion comments were taken from a transcript of the workshop. The workshop did not attempt to establish any conclusions or recommendations about needs and future directions, focusing instead on individual problems and challenges identified by the speakers. By providing an opportunity for leaders in each of the areas to share their experience and vision, we intended that the other workshop participants—as well as readers of this proceedings volume—would be able to identify new and useful ways of creating and maintaining effective research partnerships. We believe that the workshop was successful in meeting this goal.

Workshop Organizing Committee
Patrick H. Windham, *Chair*
Paul S. Anderson
William S. Millman
Gary W. Poehlein
Hratch G. Semerjian
Kathleen C. Taylor
Matthew V. Tirrell
Francis A. Via
Thomas A. Manuel (Consultant)

Acknowledgment of Reviewers

This report has been reviewed by individuals chosen for their diverse perspectives and technical expertise, in accordance with procedures approved by the National Research Council's (NRC's) Report Review Committee. The purpose of this independent review is to provide candid and critical comments that will assist the authors and the NRC in making the published report as sound as possible and to ensure that the report meets institutional standards for objectivity, evidence, and responsiveness to the study charge. The contents of the review comments and draft manuscript remain confidential to protect the integrity of the deliberative process. We wish to thank the following individuals for their participation in the review of this report:

Cheryl A. Fragiadakis, Lawrence Berkeley National Laboratory,
Barbara J. Garrison, Pennsylvania State University,
Louis C. Glasgow, DuPont,
John R. Kreick, Sanders, a Lockheed Martin Company (retired), and
Todd R. La Porte, University of California, Berkeley.

Although the individuals listed above have provided many constructive comments and suggestions, responsibility for the final content of this report rests solely with the authoring group and the NRC.

Contents

Summary

The third workshop of the Chemical Sciences Roundtable, "Research Teams and Partnerships: Trends in the Chemical Sciences," was held in Irvine, California, on May 2–3, 1999. The presentations and discussions at the workshop considered the current status of research partnerships in the chemical sciences and methods to improve the ability to form and maximize such collaborations. This volume presents the results of that workshop.

OVERVIEW

Several of the presentations provided case studies in research partnerships, some from a university view (see Wakeham, Evans), others from industry (see Tao, Kohlbrand, Sloane), and still others from the federal government (see Jackson, Powell). In addition, various issues were raised that highlight the changing nature of research teams and the environment in which they are formed. Further elaboration on each of these points can be found in the accompanying chapters.

- Companies are now buying research and development (R&D) services on a global basis. They are looking for the best deals in terms of technology fit, value, and administrative ease (including easy-to-negotiate intellectual property arrangements). Universities in the European Union now actively seek funding from U.S. corporations, and U.S. universities are beginning to wake up to this new competition. Contrasting views of major industrial funding are offered in the contributions by Wakeham, La Porte, and Evans and Tirrell.
- Since the enactment of the Patent and Trademark Amendments of 1980 (popularly known as the Bayh–Dole Act), universities retain legal title to patents on federally funded inventions. As a matter of practice, universities also often seek to retain title to any inventions developed on their campuses with industrial funds. Some universities have found that licensing such inventions can produce large amounts of money. As discussed in Chapter 1 (Mowery), most royalties collected by universities come from life sciences patents, and it may be unrealistic for universities to expect large, routine royalties in other areas of technology. For many universities, royalties may not be enough to cover even the cost of operating a technology transfer office.

• The motives and expectations of organizations appear to be changing. Faced with both rapidly changing, complex technology and cuts in internal corporate R&D, companies now emphasize external research much more than in the past. However, they also tend to expect concrete results in their R&D work with universities and federal laboratories. Universities seek corporate funding, both to increase money for research and to gain royalties. For universities, one new issue is whether private firms hold key information that the schools need for their research and education (e.g., proprietary genomics information). Another issue is whether universities now need collaborations with industry in order to draw the best faculty and graduate students. Research groups (outside of the defense area) in federal laboratories often face two motivations: the need to work with outside groups in order to stay on top of fast-changing science and technology and the need to draw corporate funds in order to maintain research capabilities threatened by budget cutbacks. Overall, these trends indicate more interest in research collaborations but also different interests that will require forthright negotiations in order to reconcile.

• Although companies and universities are actively pursuing research collaborations, there is the danger that large, exclusive arrangements at public universities may lead to concerns, on the part of both other faculty and the general public, about the privatization of public institutions. In particular, will universities be seen as responding more to specific corporate interests rather than to a range of companies or to the overall public interest? The question arises of what kind of collaborations run the risk of generating a significant political backlash.

HISTORICAL CONTEXT

In the first session of the workshop, David C. Mowery (University of California, Berkeley) established the historical context by providing an overview of three cases of research collaboration in the chemical sciences. The first case examined was the pre-World War II relationship between the Massachusetts Institute of Technology and Standard Oil of New Jersey, of which there were also analogs at other universities and other corporations. The development of fluidized bed catalysis was one significant result of this collaboration. The characteristics of this case were that personnel exchange was of central importance, technology transfer was bidirectional, academic research benefited from access to industrial facilities, industrial collaborators obtained ownership of intellectual property, and much but not all of the results of academics' work in industry was published. The second case considered was the post-World War II activities of the Research Corporation, which had by then established a reputation since its founding in 1912 for expertise in patent management and licensing. The patent licensing goals, originally philanthropic, had by then shifted to production of income which created tension with some clients' interest in using licensing to promote a broader set of objectives. The third case summarized growth of university patenting and licensing since the passage of the Bayh–Dole Act in 1980. Although many universities entered into such activities with the expectation of making a profit, it has often been the case that the most valuable product has been other non-income-producing channels for research collaboration and training of students.

Christopher T. Hill (George Mason University) presented a second facet of the historical context by reviewing various models for organized cooperative R&D between corporations and universities. Beginning with the pre-World War II era of organized industrial R&D, he traced the emergence and linkage of university research with graduate education and, as a consequence of wartime mobilization, the increased government role in support of defense. Although academic engineering education was close to practice during this period, their paths began to diverge as federal support of basic research increased after World War II. Hill then traced events that have led by steps to reintegration of university–industry cooperative R&D following the oil embargo of the mid-1970s, the increased global competition of the late 1980s, and the restructuring of industry in the early 1990s. Each of these periods saw major experiments aimed at increasing university–industry collaboration, both horizontally among partners and vertically in which research knowledge and trained students are a part of the supply chain.

William A. Wakeham (Imperial College of Science, Technology, and Medicine, London) described a university–industry strategic alliance that is being pursued at Imperial College. The converging forces that shape the strategy include a decrease in government direct funding of university income, emphasis on collaboration between industry and universities, increased student access, the desire to see research results used, and identification of new markets with special emphasis on the biomedical field. Government initiatives taken to address these issues include focusing research funding in priority areas, supporting precompetitive research in areas of commercial potential, rewarding researchers who have raised industrial funding, providing venture capital to help form spin-off companies, and setting intellectual property policies that retain faculty. Wakeham described approaches taken at his university with firms that are long term, based on trust, open-ended, customized and flexible, and multidisciplinary. These efforts are in turn having an influence on the campus through interdisciplinary research projects, targeted scholarships, new education programs, staff and student interchange, shared facilities, support of academic posts, and inclusion of small companies that cannot afford in-house research.

TEAMS AND PARTNERSHIPS IN THE UNIVERSITY SETTING

Matthew V. Tirrell (University of Minnesota), speaking for D. Fennell Evans (of the same university), described the value of research teams within universities, using the University of Minnesota's Center for Interfacial Engineering (CIE) as an example. The CIE is a National Science Foundation-supported Engineering Research Center, established in October 1988. It is chartered to carry out research, technology transfer, and education. It is a cross-disciplinary facility, intended not only to help tackle research questions of interest to industry but also to help create a new field and train students for that field. One interesting feature is that the CIE has created a new mode of collaboration among faculty. Instead of the traditional approach of an individual professor asking colleagues for advice when he or she experiences a problem, the CIE encourages, and funds, joint projects and thus promotes new research ideas. CIE also has another important feature: Because all member companies may license any patented technology developed at the center, there is relatively little patenting and licensing. Instead, Industrial Fellows from member companies have become the main mechanism for transferring CIE knowledge and technology to the companies. This fits with the center's emphasis on fostering interaction and teamwork among researchers.

John C. Tao (Air Products and Chemicals, Inc.) presented a corporate perspective on how to build university–industry partnerships. He described how his company and others now fund more external R&D than in earlier years, largely because today no one firm has all of the R&D expertise needed to deal with such factors as faster development cycle times and increasing complexity. In the case of Air Products, 4 to 7 percent of its annual R&D budgets are now spent externally. Today companies work with a wide range of R&D partners, including universities both in the United States and globally, national laboratories, other large firms, and start-up companies. In the case of industry–university partnerships, there are problems rooted in different values. Companies are very concerned about cost and timeliness. Universities focus on the advancement of knowledge and education and less on timeliness. Depending on how important timeliness and intellectual property protection are in a specific case, companies can interact with universities in various ways: gifts, use of university consultants, consortia, contracts, and hiring graduates. Consultants, contracts, and hiring graduates provide greater intellectual property protection and timeliness for a firm than either gifts or consortia. Several factors contribute to a good university–industry partnership, including a good historical relationship, complementary strengths, clear goals and roles, good teamwork and communication, good science, and clear agreement on who owns intellectual property. Tao sees his company moving toward long-term strategic relationships with a few universities, plus a number of smaller partnerships between individual Air Products' researchers and university investigators.

Todd R. La Porte (University of California, Berkeley) presented what he called a "cautionary tale" about the new, large agreement between the life sciences company Novartis and the College of Natural Resources at the University of California, Berkeley. La Porte chaired the Berkeley faculty senate's Committee on Research and thus reviewed the proposed project. In 1996, the College of Natural Resources faced two problems: old facilities with declining state support to modernize them and a new problem, lack of access to corporate-controlled information on agricultural genomes. The dean then solicited proposals from industry, of which one from Novartis was the best. However, the Berkeley faculty senate was surprised and taken aback by the scale of the proposed collaboration. To tie an entire department, and its intellectual property, so closely to one firm raised troubling questions for a publicly funded university that historically has served all of the people of the State of California. How far did this agreement put the campus down the road to privatization? The faculty senate then posed a series of questions to campus administrators on such issues as the effects of the agreement on publication, students, and California. In return, the senate received what it considered evasive answers. The senate ultimately decided not to block the proposal but to monitor it closely. This case raises issues that need thoughtful consideration.

PARTNERSHIPS WITHIN FEDERAL LABORATORIES

Henry T. Kohlbrand (Dow Chemical Company) discussed what industry seeks from federal laboratories and offered some examples of industry–laboratory collaborations. Industry R&D priorities are changing with globalization, industry consolidation, and the increased importance of information. R&D collaborations with external partners can offer speed, depth, risk sharing, and other benefits, but to be effective external technology programs require, among other things, clear priorities and an emphasis on win–win solutions. As part of Dow's external technology activities, the company has had 50 collaborations over the past five years with federal laboratories, with 14 agreements active today. One successful collaboration is the Multiphase Fluid Dynamics Research Consortium, which involves several companies, universities, and Department of Energy laboratories. This and other collaborations offer several lessons, including that people exchange is what really facilitates technology transfer, the collaboration must be a win–win for all parties, the collaboration must be cost effective, and managing intellectual property can be a significant barrier.

Nancy B. Jackson (Sandia National Laboratories) addressed catalysis as a case study in research teaming and partnerships. Sandia, being an engineering laboratory and corporate run, has a culture with a strong commitment to partnerships with industry. In 1998, Sandia had some 300 cooperative research and development agreements (CRADAs) with industry and other research partners. Today, Sandia has a particular interest in precompetitive research in areas of joint interest to both companies and the laboratory. From the laboratory's viewpoint, one motivation is the decline in federal funding in many research areas and the interest in finding outside dollars so as to maintain Sandia's capabilities. From an industry's point of view, many corporate labs have closed. Collaboration helps both sides. However, the Europeans have more experience in building industry–government collaborations, and as Jackson tries to build a virtual, multidisciplinary catalysis research effort in the United States, she finds European research centers to be the main competition. Nonetheless, U.S. federal laboratories are now motivated to seek partnerships and will continue to do so.

R&D ALLIANCES AND CONSORTIA IN INDUSTRIAL SETTINGS

Lura J. Powell, director of the National Institute of Standards and Technology (NIST) Advanced Technology Program (ATP), presented NIST as the only federal agency dedicated to industrial partnerships from its inception, and summarized a set of programs that have evolved to meet different industry needs in appropriate ways. NIST has a broad vision of collaboration as a multilateral, multimechanism activity, which will be increasingly demanded by the fast-paced, multidimensional, changing environment. The increasing requirement for multidisciplinary approaches will generate more partners in a given program. The presence of the government not only brings resources but also ensures recognition of the public interest in broadly enabling technologies. Among the NIST mechanisms are CRADAs, small business support throughout the 66 local centers of the Manufacturing Extension Program, the cooperative centers in Colorado and Maryland in the Joint Institute Laboratory for Astrophysics and the Center for Advanced Research in Biotechnology programs, and the ATP.

The ATP employs cost sharing to bridge the risk gap in developing technology between the basic science and demonstration of feasibility leading to private sector commercialization. It targets innovative technology having broad enabling consequences and is open to all areas of technology. In its tenth year, the ATP has funded 431 programs for $2.8 billion, with significant participation by universities and small businesses. A recently released report[1] confirms that the first 38 completed projects have had a significant fraction of successes, as well as some failures. All goals of the program are being attained—not only technical success, commercialization, and financial payback, but also stimulation of collaboration. Several success stories were reviewed, including the National Center for Manufacturing Science printed wiring board and the auto body consortium. Both have already had significant success in improving the competitiveness of major U.S. manufacturing industries. It is expected that only three of the projects will yield benefits equal to the total program cost since inception. A list of partnership challenges has been culled from the experience, headed by bringing the right people together and just getting started. NIST intends to look upstream for collaboration with the basic science enterprise in order to identify opportunities for ATP programs. For this audience it was noted that relatively few of the projects to date involve the chemical industry, although chemical science is integral to several of them.

Discussion confirmed that the perception of the ATP as "corporate welfare" is dead, and the debate is now a healthy one about process improvement and funding and growth levels. Although the ATP does not directly receive a funding stream from successful projects, the payback is real in terms of jobs, economic growth, and quality of life. The audience again commented on the chemical industry's historic aversion to collaboration and speculated on the reasons for it.

Christine S. Sloane, (General Motors) director of technical programs for the Partnership for New Generation Vehicles (PNGV), summarized the content and learning of this program, which is much larger than is common in the chemical industry. PNGV had its origins in concern for global warming and the desire for the transportation sector to make a step change in energy efficiency. The scope of the program requires a very high level of integrated technical and business planning and coordination,

---

[1]W.F. Long, *Performance of Completed Projects: Status Report Number 1*, NIST Special Publication 950-1 (Washington, D.C.: U.S. Government Printing Office, 1999).

involving not only the major auto manufacturers but also many suppliers and government labs. The program has three major goals: a vehicle with three times today's fuel efficiency, near-term spin-offs to standard vehicles, and major improvements in manufacturing costs and development times. All aspects of the vehicle require significant improvement—often abandoning old paradigms—to meet the overall goal. The criteria evolve over time, with the early emphasis on safety, economy, and emissions growing to a full set including marketability and commercial viability. Despite early reactions that the goal was impossible, it is likely that the concept of vehicles that get up to 80 miles per gallon will be revealed on schedule in early 2000.

Examination of the program reveals that many of the highly challenging technical problems are chemical in nature (e.g., lightweight materials, assembly, combustion, and batteries). The key technology directions were chosen in 1997, but in many cases, such as combustion-ignition engines and fuel cells, fundamental chemical understanding is lacking, so that basic research must continue in parallel with development.

In addition to the revolutionary technical challenges, complexity has been added by the inadequacy of communication among government agencies, which required the auto manufacturers to provide the linkage. Key factors for success in such a large partnership include having a significant societal goal, overcoming adversarial histories, and resolving different value schemes. Many relationships between competitors, suppliers and customers, and manufacturers and government had to be reinvented. The consensus is emerging that economic competitiveness and environmental benefits go together. Delivery as promised is a key to credibility. The commitment must be commensurate with the goal, in this case at the CEO level. It is crucial to recognize the effect of changing market forces and the need for customer acceptance—for instance, fuel economy was second among consumer references in 1980 but fifteenth in 1999.

Emerging realities of the partnership are that government is not monolithic, delay and budget creep must be expected, Congress cannot be expected to understand science and technology nor scientists and engineers to understand Washington, and agency missions are variable but important. Despite the problems, the program is on the route to success and delivering many benefits.

Discussion clarified that issues raised by the merger of Chrysler and Daimler were favorably resolved, because the new entity still conducts R&D and manufactures in the United States. Despite its origins, both political parties have now endorsed the program, so political risks are not severe. The public taste for gas-guzzling sport utility vehicles has to be considered, but the technologies under development can and will be transferred to larger vehicles. No single technology area is emerging as the primary key; all are essential. The Japanese pursuit of similar goals has stimulated Detroit's competitive spirit, especially because it must be kept in mind that the relevant market is not solely the United States but the world. It is not possible to say with certainty that PNGV has permanently altered the level of trust between General Motors and the federal government.

1

The Evolving Structure of University–Industry Collaboration in the United States: Three Cases

David C. Mowery
University of California, Berkeley

INTRODUCTION

Collaborative relationships between U.S. universities and industry have changed significantly over the past 60 years, as typified by three cases of research collaborations in the chemical sciences. I begin with a summary of the development of the discipline of chemical engineering in the 1920s and 1930s, with a particular emphasis on the role of the Massachusetts Institute of Technology (MIT) and its collaboration with industry. The second case discusses management of patenting and licensing by universities in the postwar period and the development of the Research Corporation, a nonprofit entity created to manage the licensing activities of U.S. research universities. The third case is the Bayh–Dole Act, which sparked another wave of change in university–industry collaboration after its passage in 1980.

One message that I want to impress upon this audience is the lengthy history of university–industry collaboration in the United States. Indeed, both the U.S. research university and the organized pursuit of research and development (R&D) in industry trace their origins back roughly 125 years, and have grown in parallel throughout the twentieth century.[1] The historic involvement of publicly funded universities in the United States with agricultural research, much of which was applied in character, and the involvement of these universities with the agricultural users of this research are well-known aspects of U.S. economic history. But throughout this century, the decentralized structure of U.S. higher education and the dependence of public and private universities on local sources of funding also meant that in a broad array of nonagricultural fields, ranging from engineering to physics and chemistry, collaborative research relationships between university faculty and industry were common.[2]

---

[1] D.C. Mowery and N. Rosenberg, *Paths of Innovation: Technological Change in 20th Century America* (New York: Cambridge University Press, 1998).

[2] N. Rosenberg and R.R. Nelson, "American Universities and Technical Advance in Industry," *Research Policy* 23: 323–348, 1994.

OLD-STYLE COLLABORATION

I label the first case "old-style collaboration," although this type of collaboration continues in many sectors of U.S. industry and in many U.S. universities. There are several key characteristics of this collaboration, many of which are illustrated by the collaboration between MIT and Standard Oil of New Jersey that contributed to the development of the discipline of chemical engineering in the United States before 1940. This academic discipline was developed with major contributions from MIT, the University of Wisconsin, and the University of Illinois. But many of the innovative collaborative relationships that underpinned the growth of chemical engineering was centered at MIT. Arthur D. Little, Warren Lewis, and other faculty encouraged the development of these collaborative relationships, involving research and teaching, the exchange of students in cooperative education, the foundation of the school of chemical engineering practice, and a nearly parallel growth of organized research in engineering at MIT and in industry. The development of chemical engineering research, teaching, and practice was influenced by the symbiotic relationship between Standard Oil of New Jersey and MIT faculty who worked to codify, advance, and disseminate the key tenets of the emergent discipline.

Much of the collaboration during this period involved joint development of these new practices in both the academic and the industrial laboratories combined with relatively widespread dissemination, particularly through teaching and textbooks. The Standard Oil refinery in Baton Rouge, Louisiana, also played a key role as an unofficial external laboratory and employer of a great many of the graduates and a number of the faculty at MIT in the school of chemical engineering.

The MIT–Standard Oil collaboration culminated in the development of fluidized bed catalysis in 1941. Research conducted at MIT complemented research done in the Baton Rouge refinery and produced an important patent that was assigned to Standard Oil. Although intellectual property and formal patents clearly were an important component and an important output of this research activity, the university's direct role in managing, licensing, and seeking the assignment of the intellectual property was quite different from what we observe today. Indeed, it contrasts with the policy that evolved at MIT during the decade after this breakthrough in catalysis.

The key to this style of collaboration was personnel exchange—primarily from MIT to industry—through faculty consulting, faculty rotations to and from industry, and placement of graduates. Personnel exchange was a very important component of technology transfer, bringing expertise from MIT to industry and transferring practical knowledge from industry back into academia, where it was refined and codified, supporting the development of a broader engineering discipline. As in many other areas of engineering or scientific research, access by faculty to industrial facilities was important, as the scale and type of equipment in industry often were not available in universities. The industrial collaborators obtained the ownership of or were assigned the intellectual property resulting from collaboration, and a great deal, although not all, of the results of the research by academics in the industrial context were published.

Many of these characteristics still apply to much of the collaboration operating between U.S. universities and industrial firms today. Nevertheless, many of these collaborations contrast with others that center on patent licensing. There are some important contrasts to keep in mind between this "old style" and what we see emerging since 1980. A number of features of the post-1980 relationship between U.S. universities and industry are illustrated by the origins, growth, and decline of the Research Corporation during the 1940–1980 period.

THE RESEARCH CORPORATION[3]

Origin and History

The second case of collaboration, that of the Research Corporation, is interesting in its own right because of its origins and its unusual function within the developing academic and industrial R&D system in the United States. It also provides insight into the characteristics of technology licensing—particularly the licensing of academic technology—by an organization that has been in place for a much longer period of time than most individual campus technology transfer or patent licensing offices.

The Research Corporation was founded in 1912[4] on the basis of a technology invented at the University of California at Berkeley, where Frederick Cottrell, a professor at the university, developed antiprecipitant pollution technology for reducing particulate emissions. His work was motivated in large part by a desire to mitigate the pollution generated by the activities of local firms in the San Francisco area. Cottrell developed a substantial portfolio of patents that he licensed, using the proceeds from the licensing activity to support scientific research through the extension of grants to other researchers. The ultimate goals of this patent licensing operation thus were philanthropic, in some contrast to more recent university licensing activities.

For the first several decades of its existence, the Research Corporation focused on expanding and marketing its patent portfolio in antipollution technology. As a result, the Research Corporation developed a reputation and expertise in patent management and licensing, accumulating knowledge of patent application procedures, management of patent litigation, and management of patent licensing. In addition, it received a number of donations of patents from academic inventors. The Williams–Waterman patent for vitamin B$_1$ in the early 1930s was one such patent given to the Research Corporation by the inventors so that the corporation could manage the licensing and use the income to support other areas of scientific research.

The Research Corporation expanded its patent management and licensing activities through an agreement with MIT in 1937. Karl Compton, president of MIT, wished to extract more income from patents obtained by MIT faculty. Based on the Research Corporation's reputation, as well as a personal relationship between Compton and the then president of the Research Corporation, an agreement was negotiated between MIT and the Research Corporation. Compton's decision to contract with a third party to manage MIT's patent portfolio was influenced by MIT professor Vannevar Bush, who was an industrial inventor of some repute with a substantial number of patents to his name. Bush believed that MIT should not be involved directly in management of patents, and believed that MIT particularly should avoid direct involvement in management of licensing contracts because of the risk that such activities would be criticized by politicians or industrial firms. The agreement committed the Research Corporation to use its best efforts to obtain patents on MIT inventions, license the patents, and, not insignificantly, to pursue infringers.

The Research Corporation's work with MIT grew moderately until about 1940, when MIT's expanding wartime research made patenting and licensing less important. Nevertheless, the Research Corporation negotiated numerous agreements with other U.S. universities after World War II. The

---

[3]D.C. Mowery and B. Sampat, "Patenting and Licensing of University Inventions: Lessons from the History of Research Corporation," paper presented at the conference on R&D and Economic Growth in the 20th Century, Haas School of Business, University of California, Berkeley, March 27, 1999.

[4]"The Research Corporation, An Experiment in Public Administration of Patent Rights," *Journal of Industrial and Engineering Chemistry* 4 (12), 1912.

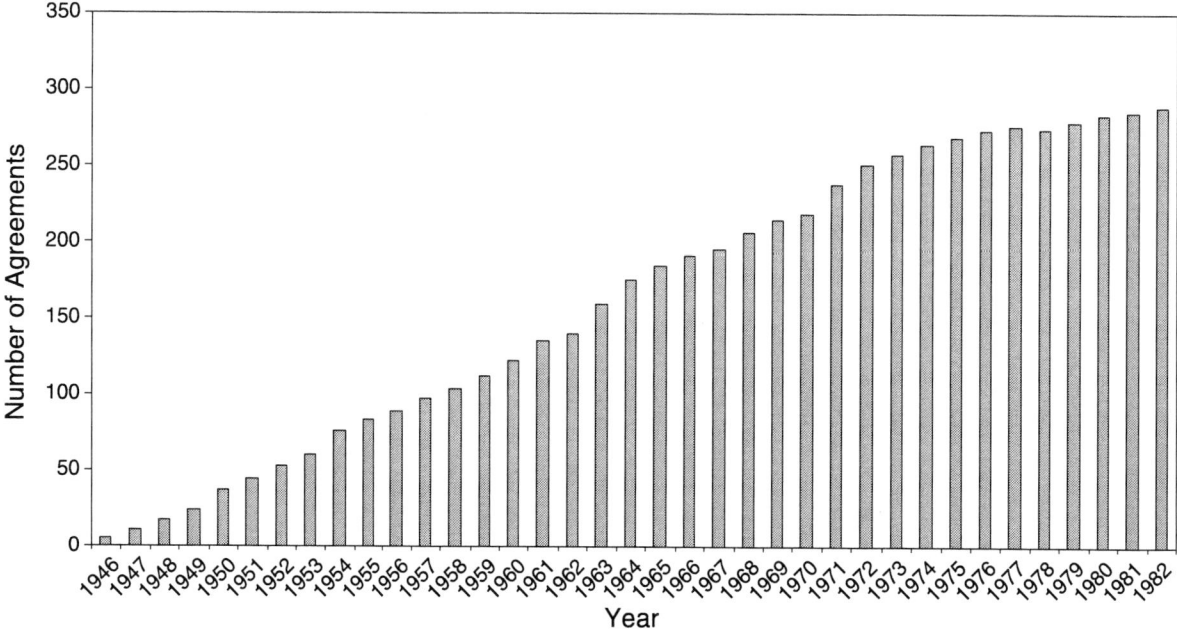

FIGURE 1.1   Invention administration agreements: 1946–1982.

number of administration agreements between the Research Corporation and individual U.S. universities grew from roughly 5 in 1946 to more than 280 by the early 1980s (see Figure 1.1). The Research Corporation sought cost savings and efficiencies through its centralized management of a substantial and diverse patent portfolio. However, these cost savings proved to be elusive; centralized management for such a large, diverse patent portfolio was not very cost effective.

Difficulties with Centralized Patent Management

The anticipated cost savings associated with centralized management of patenting and licensing were not realized for several reasons. First, the expertise developed through experience in patenting and licensing in one area of technology did not always improve performance in these activities in other technology fields. The Research Corporation focused its efforts and became most proficient in patenting and licensing in the biomedical area, as demonstrated by a number of important vitamin patents and some early drug patents. Indeed, MIT became frustrated with the Research Corporation's lack of expertise in patenting and licensing electronics inventions, a field in which MIT accumulated numerous patents from its wartime R&D. In 1947, MIT terminated its exclusive relationship with the Research Corporation as manager of its patent portfolio and used other organizations—such as large patent law firms—to supplement its relationship with the Research Corporation.

Adding to the costs of the licensing process was the need for frequent interaction between the Research Corporation representatives and client universities. The universities with whom the Research Corporation developed patent licensing agreements insisted on a great deal of "handholding." Geographic proximity was essential to the Research Corporation's relationships with a large number of geographically disbursed institutions. The need for geographic proximity had not been foreseen by the

Research Corporation, which increased its costs because considerable travel expenses and a large number of branch offices were necessary, and it made it very difficult to reap the efficiencies associated with this centralized management.

Finally, conflicts developed from the Research Corporation's contractual agreement with many of its institutional clients to extract as much income as possible from patent licensing. In some cases, the client universities found that the Research Corporation's focus on income maximization impaired the universities' relationships with industrial firms that supported research or were important philanthropic academic donors. The MIT patent on magnetic memory for computer technology was one such case. In this case, the efforts of the Research Corporation to prosecute the patent and to litigate the patent against alleged infringement by IBM produced quite a bit of tension with a major corporate supporter of the university. MIT took back this patent from the Research Corporation and licensed it on more generous terms to IBM. This conflict between the goal of maximizing income and a broader set of institutional goals of the universities remains significant in the operation of a number of contemporary technology transfer and patent licensing offices in the United States.

The overall costs of the Research Corporation and the cost per patent rose substantially from 1960 to 1982 (see Figure 1.2). As a result, the Research Corporation encountered cost pressures that in turn caused a decline in its net income (see Figure 1.3). The mid-1970s in particular brought recurrent deficits. These deficits appeared prior to the Bayh–Dole Act, which led a number of universities to enter technology licensing independently of the Research Corporation.

Another important aspect of the Research Corporation's activities, which applies to most contemporary licensing offices, was the dominance of its overall income by a very small number of inventions,

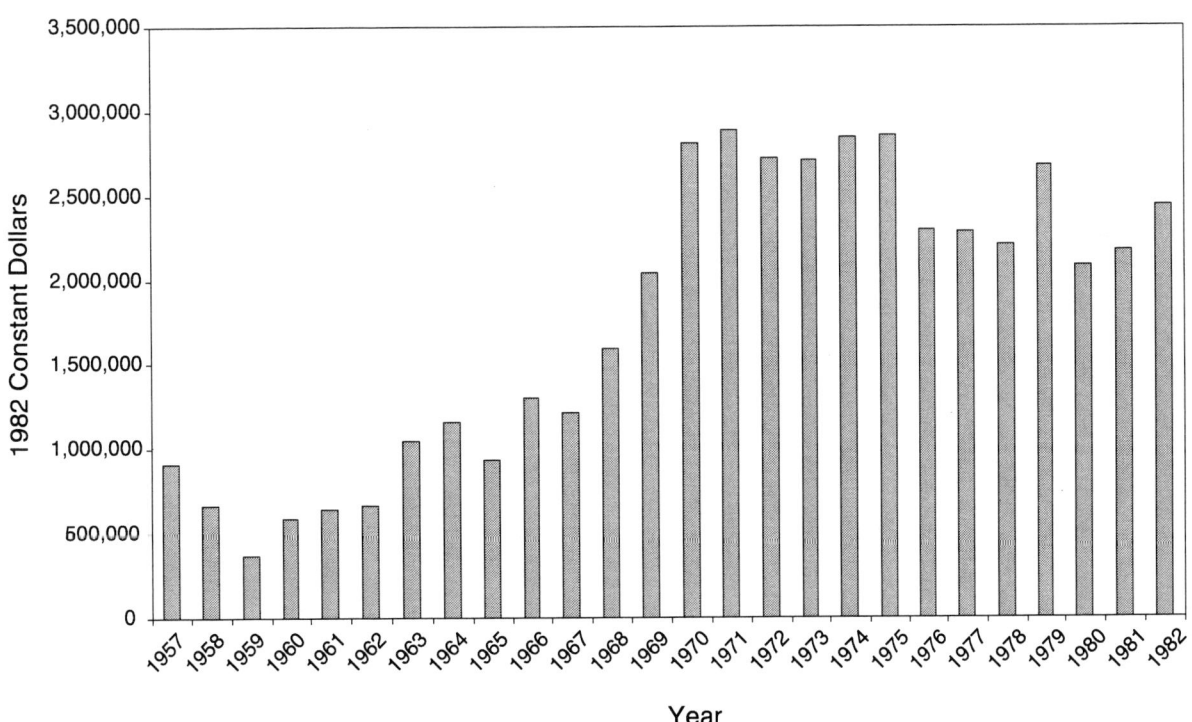

FIGURE 1.2   Costs of the patent development division, 1957–1982.

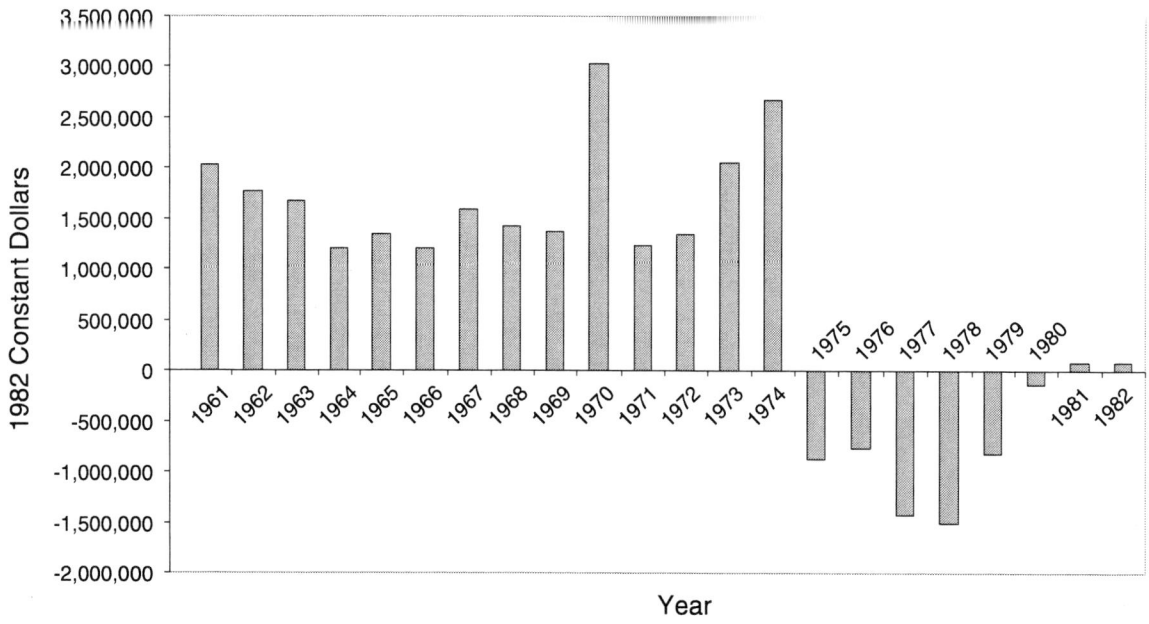

FIGURE 1.3   Net income of the Research Corporation, 1961–1982.

and the dominance of those inventions by biomedical and life sciences patents. From 1960 to 1975, 75–90 percent of the Research Corporation's income was provided by patents on only four inventions: nystatin, cortisone, reserpine, and hybrid corn. Along with the growth of costs and the associated emergence of operating income deficits, the expiration of these "home-run" patents in the mid-1970s intensified financial pressures on the Research Corporation.

In summary, the history of the Research Corporation suggests that centralized management of patenting for multiple institutions is difficult and is not always as cost efficient as frequently suggested. This is a consequence of the limited spillovers that often exist of management expertise across technology categories. This tension between centralization and decentralization characterizes a number of contemporary university licensing offices. In addition, the conflict between the income-maximizing technology broker and the interests of institutional clients cannot be neglected. And finally, any analysis should recognize the concentration of licensing income among a very small number of inventions.

BAYH–DOLE[5]

Expanding Patenting and Licensing Activity

The third of these cases is the so-called Bayh–Dole era. The Bayh–Dole Act was passed in 1980 and facilitated the licensing of inventions based on patents from federally funded research by U.S. universities

[5]D.C. Mowery, R.R. Nelson, B. Sampat, and A.A. Ziedonis, "The Effects of the Bayh–Dole Act on U.S. University Research and Technology Transfer: An Analysis of Data from Columbia University, University of California and Stanford University," forthcoming in *Industrializing Knowledge*, L. Branscomb and R. Florida, eds. (MIT Press, Cambridge, Mass., 1999).

and other nonprofit organizations. It codified and rationalized federal policy, which previously had been a patchwork of agency-specific policies and agreements negotiated with individual universities. The Bayh–Dole Act did not make licensing possible that had previously been impossible, but it stated that Congress, and thus the federal government, believed that this licensing was a legitimate use of federal funds and mandated a common government-wide policy.

The act produced rapid expansion of U.S. universities' patenting and licensing activities. The number of licensing offices maintained by U.S. universities has grown from roughly 25 in 1979 to well over 200 today. The number of patents assigned to U.S. universities has more than quadrupled during that period of time.

The research summarized in Mowery et al. (1999) examines data from Stanford University, Columbia University, and the University of California (UC) system before and after Bayh–Dole. This includes two universities, the UC system and Stanford University, that were active in patenting and licensing before Bayh–Dole, and Columbia University, which has become a major licensor of technology since the passage of Bayh–Dole. There was increased patenting activity after Bayh–Dole at both UC and Stanford, but there was also an increase in university patenting at both institutions substantially before the passage of the Bayh–Dole Act. The increased patenting activity was concentrated in the biomedical area. During the 1975–1990 period, the share of biomedical patents in UC patents increased from 25 percent to 65 percent. And the major increase in the biomedical portion of UC inventive output took place well before the passage of Bayh–Dole. We find a similar timing and composition of expanded inventive output at Stanford (see Figure 1.4). More than Bayh–Dole underpins these developments, something that must be kept in mind in any evaluation of this law's effects.

The increase in biomedical invention disclosures and patenting activity prior to 1980 appears to be a consequence of the rapid growth of federal funding in biomedical research, notably under the auspices of the National Institutes of Health, and especially the war on cancer that began in the early 1970s. These events triggered a wave of scientific advances in molecular biology and led to the development of the biotechnology industry. For example, the application for the Cohen–Boyer patent predates Bayh–Dole. Changes in federal policy toward intellectual property rights in the late 1970s and 1980s also enhanced the legal strength and economic value of patents in the biomedical and biotechnology areas.

Table 1.1 depicts trends from 1970 to 1995 in gross income from licensing (in constant 1992 dollars) at UC, Stanford, and Columbia. The UC earned slightly more than $1 million in FY 1970, and almost $60 million in FY 1995. Stanford University's earnings increased from $180,000 in FY 1970, the first year of operation of Stanford's Office of Technology Licensing, to almost $36 million in FY 1995. Columbia University, a more recent entrant into the patenting arena, did not even have a technology licensing office until 1982. Columbia earned $540,000 in FY 1985, and this grew to nearly $32 million in constant dollars by the mid-1990s. Equally important is the domination of these income flows at all three universities by a small number of inventions. The share of gross income from the top five inventions at the UC was almost 80 percent in 1970, although by 1995, this figure had dropped to 66 percent. The similar share at Stanford was 69 percent of gross income in 1975, which increased to 85 percent in 1995. Columbia's income has been even more concentrated, with more than 90 percent of gross income for FY 1985–1995 flowing from the five most lucrative licensed inventions.

These "top five" inventions at each university also are derived largely from biomedical technologies. At UC, because of the early importance of agricultural inventions, biomedical inventions account for only 34 percent of its licensing income from the five most lucrative inventions in 1970. But by the end of the period, all of the top five UC inventions were biomedical. For Stanford, the biomedical inventions accounted for 87 percent of the earnings of the top five inventions in FY 1975, a figure that

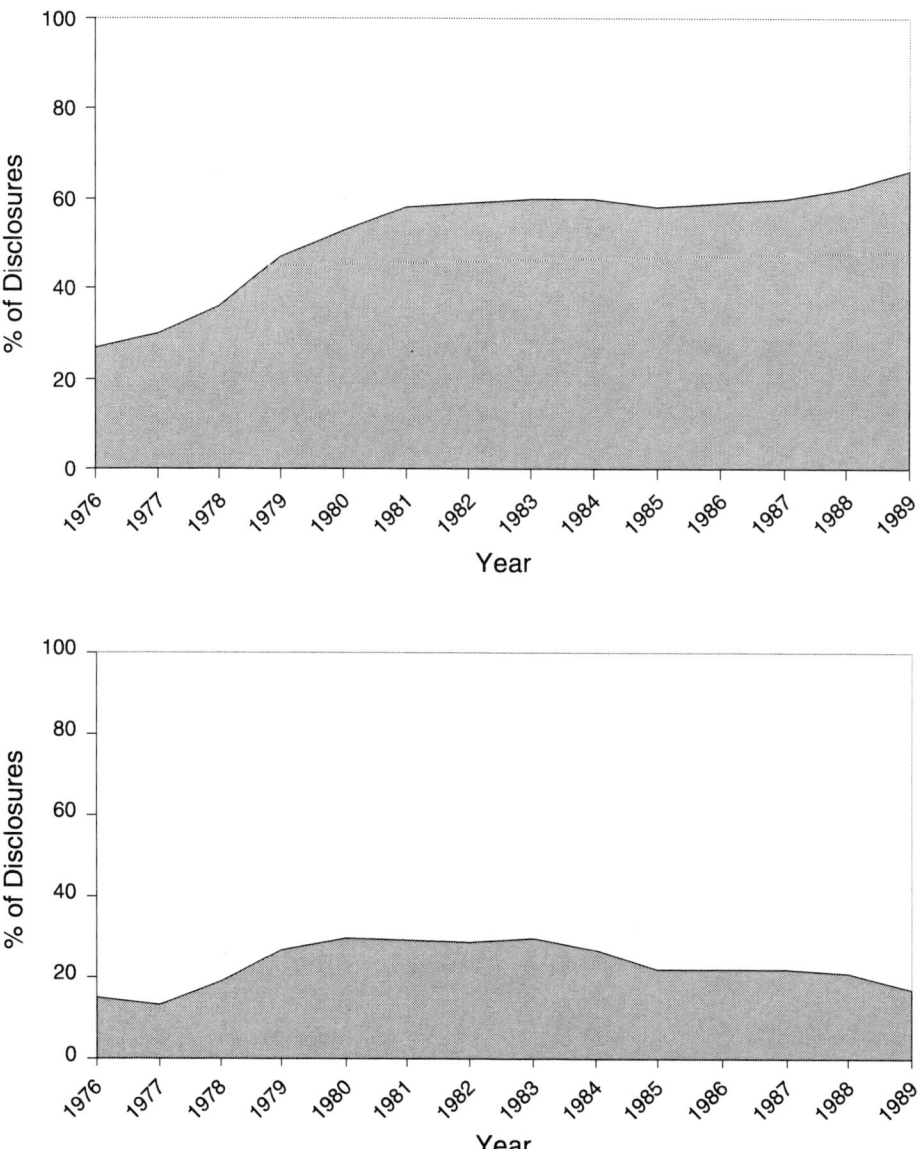

FIGURE 1.4   Biomedical disclosures from 1976–1989 at the University of California (top panel) and Stanford University (bottom panel).

increased to 97 percent in 1995. At Columbia, the biomedical inventions' share of the income from the top five inventions grew from 81 percent in FY 1985 to 91 percent in FY 1995.

Just as was true of the Research Corporation's licensing income, these three research university systems derive the majority of their licensing income from biomedical inventions. The prominent role of biomedical inventions in licensing income can be attributed to a number of factors, among the most important of which is the strength and value of intellectual property in this technology. Biomedical

TABLE 1.1 Selected Data on University of California, Stanford University, and Columbia University Licensing Income, FY 1970 to 1995 (all earnings are given in thousands of 1992 dollars)

University and Selected Data	1970	1975ᵃ	1980	1985	1990	1995
University of California						
Gross income	1,140.4	1,470.7	2,113.9	3,914.3	13,240.4	58,556.0
Gross income of top five earners	899.9	1,074.8	1,083.0	1,855.0	7,229.8	38,665.6
Share of gross income from top five earners	79%	73%	51%	47%	55%	66%
Share of income from top five earners associated with biomedical inventions	34%	19%	54%	40%	91%	100%
Share of income from top five earners associated with agricultural inventions	57%	70%	46%	60%	9%	0%
Stanford University						
Gross income	180.4	842.6	1,084.4	4,890.9	14,757.5	35,833.1
Gross income of top five earners		579.3	937.7	3,360.9	11,202.7	30,285.4
Share of gross income from top five earners		69%	86%	69%	76%	85%
Share of income from top five earners associated with biomedical inventions		87%	40%	64%	84%	97%
Columbia University						
Gross income				542.0	6,903.5	31,790.3
Gross income of top five earners				535.6	6,366.7	29,935.8
Share of gross income from top five earners				99%	92%	94%
Share of income from top five earners associated with biomedical inventions				81%	87%	91%

ᵃ Figures for Stanford University are for FY 1976 instead of 1975.

patents are profitable because intellectual property protection is strong and the economic significance of an individual patent is substantial. That is, a patent on a molecule for a pharmaceutical company is typically both stronger and more valuable than a patent on a design technique for an integrated circuit.

The data on the growing number of university licensing offices in the United States indicate substantial entry into patenting and licensing by universities as a result of Bayh–Dole. But for many of the universities that have entered the pursuit of licensing, the lack of home runs and a relatively modest flow of biomedical inventions in particular mean that their licensing activities are unprofitable. Although profit is by no means the only or necessarily the most important motive for establishing a licensing office, the probability that a given university will realize a net profit on its licensing activities appears to be modest at best.

A second issue raised by Bayh–Dole involves its effects on collaboration between U.S. universities and industry. The emphasis on intellectual property for both the university and the potential industrial partners in collaborations that has been sparked by Bayh–Dole can either support or limit such collaboration, and these effects will differ among industries and among segments within industries. In an industry as large and as diverse as the chemical industry, for example, some firms find that the emphasis on patenting and licensing facilitates collaboration. Other firms in other segments of the industry, such as heavy chemicals, find that this emphasis on defining at the outset the ownership of intellectual property flowing from collaboration is an impediment to collaboration.

Assessing Bayh–Dole

Are the effects of Bayh–Dole on U.S. universities and industry desirable? We believe they are mixed. We find little evidence that the incentives created by Bayh–Dole for patenting, licensing, and associated transfer of technology have shifted the academic research agenda toward applied research, a concern voiced by some observers. But this is a very tentative and preliminary conclusion.

In some spheres, however, such as biotechnology research tools, including research materials or genetic material, the invention is itself an input to the scientific research process. If universities expand their patenting and licensing of these research tools, research results formerly placed in the public domain will be taxed by their inventors. The Cohen–Boyer patent, for example, embodies knowledge that might have been published and widely exploited by industry without a patent. Cohen–Boyer was licensed on a nonexclusive basis, so it was widely employed. But it is not clear that the transfer and application of this knowledge was actually facilitated by patenting and the presence of the university as a licensor. Moreover, in the burgeoning area of research tools, a proliferation of patents and restrictive licensing agreements could limit the use of important inputs into scientific research. There is also a broader issue of whether the establishment of private property rights over publicly funded research is itself desirable.

CONCLUSIONS

The three cases discussed here support several conclusions about the origins and structure of university–industry collaboration. First, the form and structure of collaboration have changed over time. In the early twentieth century, there was a complementary relationship between the emerging industrial research activities of private firms and the university-based R&D. Gradually, MIT and other U.S. universities sought greater control of the management and the income associated with university-developed intellectual property. By the 1980s, the structure of many university–industry collaborations was precisely what Vannevar Bush had sought to avoid, as universities were directly involved in patenting and licensing research results. It remains an open question as to whether Bush's concerns over the potential exposure of universities to political criticism resulting from their direct involvement in patent management were well founded.

These changes in the structure of the relationships between U.S. universities and private firms have had other consequences. The complementary relationships that developed in some areas of industry and technology have been replaced or supplemented by a more directly competitive relationship between industrial and academic research in some areas. Recent agreements such as that between Novartis and the University of California, Berkeley (see Chapter 6 by Todd La Porte), align a group of faculty in a university with a single firm, affecting the competitive position of that firm vis-à-vis others. In the area of research tools, universities often compete with and rely on biotechnology firms in developing and licensing research tools and genetic material. In many cases, universities need the research material from the biotechnology companies, but also compete to license these research tools to pharmaceutical firms, among others.

There is now a very complex mix of competition and cooperation. Consider Genentech, a firm that came out of the University of California, San Francisco, and an early licensee of the Cohen–Boyer patent. It is now entangled in a very interesting web of litigation with the UC. Biotechnology is unique in many respects and does not describe the entire landscape of university–industry collaboration, but at least in these instances, the relationship is starting to look very different from the "old-style" collaboration and has taken on elements of competition in place of cooperation. The effects of the emergent competitive relationship between university and industry research are very unclear for political support for the continued public funding of the university research.

The lengthy and rich history of university collaboration in the United States is an outgrowth of some unusual structural characteristics of the U.S. system of higher education. By comparison with those of Japan, Germany, France, and the United Kingdom, the U.S. higher education system has been much larger throughout this century. Almost any international comparative analysis shows that the number of students and the number of institutions are much greater than in these other industrial economies. The U.S. system also is characterized by a more diverse mix of institutions, including research universities, liberal arts colleges, and public and private institutions. These diverse institutions are not managed nationally or centrally, but compete fiercely with one another for prestige, for students, for faculty, and for resources. This is a very different structure from that of most other industrial economies.

These differences have direct implications for the effects of foreign emulation of the Bayh–Dole model. A number of foreign governments, such as that of Japan, have passed legislation that creates a legal basis for so-called "technology licensing organizations." But the radical differences between the Japanese and the U.S. university systems mean that the effects of such legislation may be quite different from those of Bayh–Dole, keeping in mind that Bayh–Dole alone does not explain the upsurge in patenting and licensing activity in the United States. The other structural characteristics of the U.S. university system, rather than Bayh–Dole alone, have contributed powerfully to the longstanding collaboration between U.S. universities and industry that dates back almost a century.

I have focused here on a relatively narrow slice of the channels through which universities in the United States transfer technology, interact with, and collaborate with industry. But it is important to recognize the number, breadth, and diversity of these channels. They include publication, training of students, and faculty consulting, in addition to these more conventionally defined channels of technology transfer. One of the remaining challenges and potential risks associated with the Bayh–Dole Act is the danger that an excessive focus on intellectual property and licensing and patenting as the primary channel for technology transfer will have a chilling effect on other channels. This is a risk, and not yet a reality, but is something that needs to be kept in mind when managing these activities within universities and when explaining these activities and relationships to other constituencies, both political and industrial.

DISCUSSION

Thomas Manuel, Council for Chemical Research, Inc.: You mentioned a common belief that a very small fraction of university patent offices are profitable. Are there valid data that attest to that? It seems to be anecdotal truth, but it's hard to find the numbers.

David Mowery: Surprisingly, these data do not exist. The surveys published by the Association of University Technology Managers, for example, report gross income, not net income. There is one piece that was published by two academics from Portland State University, and whose accuracy I tend to discount, that suggests that the proportion of profitable offices could be as high as 30–40 percent. This is probably a high estimate. But that is the only estimate I have seen. My guess is that the true proportion is probably closer to 20 percent.

Francis Via, General Electric: Thank you for a wonderful overview, especially the quantitative data of trends we all believe in. As a result of these data, have you seen cases in which universities are changing their role relative to interacting with particular industries, such as the steel industry or the chemical industry?

Being associated with the chemical industry, I have concerns that cooperative programs are going overseas as a result of the drive of intellectual property rights or becoming global, and we are becoming global, but to a greater extent than would be justified otherwise. And many times the expectations of universities are associated with the biotechnology.

Once we did some research in this area for the Industrial Research Institute, in which we looked at the 50 top income producers for all universities in 1994. And of the top 50, only one was associated with chemistry—separations technology. The proliferation of this information is very important to us. But are you seeing changes in universities interacting with various industries in that there may be greater flexibility with one industry that is more interested in the other approaches to technology transfer rather than licensing?

David Mowery: This is a good question, and I do not have an easy answer to it. Some universities have expressed concern that a focus on income, on licensing, and licensing with a view to maximizing income will have a negative effect on relationships with firms in other sectors such as nonbiotechnology. But there is a high-level administrative pronouncement to this effect, combined with a set of internal norms for evaluation and for rewarding the technology managers that continue to focus heavily on income because, of course, it's what can be counted. It's what you can see.

This broader array of objectives is in many cases difficult to measure, difficult to quantify, operates over a longer time horizon and, therefore, it's much more difficult to get it into the managerial consciousness of the people in the technology licensing office.

But it's a risk, and you are exactly right. We do see the growth of "foreign competition" in the demand for their services, for example, the ability of Procter & Gamble to establish relationships with universities and the People's Republic of China without as much concern over intellectual property rights. U.S. universities are aware of this and are so scared of it that they cannot even talk about it. So it's an issue that is not very clearly recognized or articulated.

John Tao, Air Products and Chemicals, Inc.: I agree with a lot of what you said in your presentation. In your answer to Thomas Manuel, I agree with your number of 20 percent. In the survey by the Association of University Technology Managers, at least the 1997 numbers, if you use the table that reports the number of professionals in the technology transfer function, apply a number including overhead, and weigh that against the gross income, most of the technology transfer offices are not profitable.

Laren Tolbert, Georgia Institute of Technology: There is another number that is even more difficult to arrive at, and that is the lost opportunity. That number would weigh all this down even further. There are any number of cases I could point to in which grants were not awarded and agreements were not signed because it was never possible to reach an intellectual property agreement. This is another issue that needs to be thrown into the mix.

Robert Lochhead, University of Southern Mississippi: The focus on biomedical innovation intrigues me, and it would be interesting to look at the deeper reason for it. One of the reasons might be that the culture of biomedical researchers differs from that of other researchers. Is this driven by more-focused government funding? Is the time ripe, for example, for understanding genetics the same way that understanding the atom and molecules drove chemistry 100 years ago? Or is it that the biomedical field is just getting more of its share of bright people today because drivers for the biomedical segment are so common?

David Mowery: Many of the points you suggest are among the drivers. Federal funding for biomedical research has increased in real terms at a rate of about 7 percent per year for approximately the past 30 years. That cannot be said about many other areas of federal funding for academic research. But there are major scientific advances that, as you suggest, have revolutionized the field and, very importantly, this is an area in which the scientific frontier is arguably closer to commercially attractive applications than many other areas of research.

And the funding and intellectual excitement attract very bright people. And finally, there is the combination of strengthening of intellectual property rights that occurred in the United States during the 1980s, in particular, and operated with particular force precisely in the life sciences and biomedical area, and this has facilitated licensing. Licensing contracts in some respects are more straightforward to draw up. The value of an individual patent is higher in many cases and the strength of that patent is greater. So all of these factors have elevated the significance of the biomedical area within this licensing. And, as has been suggested several times, the danger is that university technology licensing managers or university presidents think that the whole world is as profitable as biotechnology and that the research and the licensing goals do not come into conflict. That, I think, is the challenge here because the money looks good when you see numbers like those for Stanford, Columbia, and the UC system.

Ashok Dhingra, Dupont: I enjoyed your insight into partnerships. My question is in regard to Procter & Gamble undertaking R&D in China and the globalization of their businesses. R&D is being globalized. General Electric, United Technology, Dupont, and many other companies are now moving into Asia Pacific to start these partnerships for many reasons, not only for capturing some of the core competencies, but also helping develop the local markets, especially the technology support that may be needed to grow the markets there. Do you have any study or any data that show the effectiveness of these partnerships in China, India, or Singapore versus the partnerships that are done in the United States?

David Mowery: No. I wish I did. It is a very interesting topic to pursue because my impression is that U.S. universities have their heads in the sand on this issue. Yet it's a reality. It would be very interesting to look at this in more detail.

2

Partnerships in Research:
The Evolution of Expectations

Christopher T. Hill
George Mason University

INTRODUCTION

In this chapter, I address the evolution of cooperation among companies and between companies and universities in R&D, discuss what participants have expected from this cooperation, and examine the roles government has played in facilitating, responding to, or driving cooperation. My core argument is that we have experienced very substantial changes in how we think about R&D and that our different ways of considering R&D have been accompanied by different expectations about R&D cooperation.

My approach is historical and chronological, considering five eras over the past 75 years. These eras are the pre-World War II era, the postwar era through the late 1960s, the post-oil-embargo era of experimentation, the competitiveness crisis era of the 1980s, and the present era of industrial restructuring. Each of these eras features different approaches to and expectations from cooperative R&D.

THE PRE-WORLD WAR II ERA

The pre-World War II era witnessed a transition from industrial technology based largely on the work of inventors like Edison, Ford, and the Wright brothers, to the development of new technology by separately organized R&D departments in large corporations. American universities were only beginning to challenge Europe in fundamental chemical research and the production of doctoral-level chemists and engineers. The government role in R&D was limited. Government arsenals did a modest amount of research in support of the Army and Navy, but it was widely agreed that the federal government should not support fundamental research in universities because its influence would be corrupting. The government did support some applied academic research in agriculture and mining.

Before the war, formal R&D cooperation between universities and industry was rare. To be sure, early leaders like Arthur D. Little at the Massachusetts Institute of Technology (MIT) and Charles F. Burgess at the University of Wisconsin built industrial chemical corporations on the foundations of their academic research, and academic engineering education was quite closely related to industrial practice.

However, an effort by Karl Compton, the president of MIT, to raise $3 million from a number of firms to support academic R&D was unsuccessful. Some attempts were made in the U.S. Congress in the mid-1930s to create a program of government support of industrial research, but these attempts were roundly rejected, not only on ideological grounds but also because influential members of Congress believed that important inventions came from the minds of individual creative men like Edison, and that organized R&D could contribute nothing useful to the furtherance of new technologies.

THE POSTWAR ERA

World War II changed many things in American society. The nation's R&D infrastructure, such as it was, was mobilized by Vannevar Bush to support the war effort. The R&D contract was invented as a means for the government to buy something it could not describe in advance—new knowledge. Major new weapons laboratories were created, and universities for the first time received large amounts of money to carry out research for the government. The newly created defense industry, based on mobilization of preexisting civilian industry, was given government funds to conduct and/or manage large-scale military R&D.

As the war wound down, R&D leaders sought ways to ensure that federal funds would continue to flow to the universities, weapons laboratories, and defense contractors. Even though the United States emerged from the war as the world's dominant technological nation, there was concern that R&D would not be sustained to help meet the challenges of reenergizing the civilian economy, addressing unmet health needs, and meeting the emerging threat from the Soviet Union. Vannevar Bush led an effort to crystallize the new views of the role and organization of R&D in his seminal report, *Science: The Endless Frontier*,[1] released in 1945.

The Bush report argued that R&D could help address important national problems, and it called for a government-supported National Research Foundation to support R&D for both civilian and military purposes. This report codified the linear model of innovation, in which investments in basic research could be expected to produce useful outcomes. The report and its supporters envisioned a system in which government would support academic research and industry would exploit the results for practical ends.

Bush lost the organizational battle—his comprehensive National Research Foundation was not adopted. The Office of Naval Research, the Atomic Energy Commission, and the National Institutes of Health were all established or expanded before a more limited National Science Foundation (NSF) was established in 1950, largely to support academic fundamental research not covered by the other agencies.

However, Bush's vision of the importance of R&D to national well-being, as well as his ideas about which institutions in society should do what kinds of R&D, did get adopted, not only in government but in industry as well. His report ushered in what I call the "classical" period or the "golden age" of R&D in America. From about 1950 to 1970, R&D infrastructure and funding grew exponentially, largely on the Bush model. Responding to the Bush model of the centrality of basic research, many major corporations set up fundamental research laboratories, usually isolated from the company's business operations.

A key aspect of the R&D model at the time was corporate self-sufficiency; that is, during the golden age, large corporations sought to develop their own capabilities in each area of technology on which

---

[1] Vannevar Bush, *Science–The Endless Frontier: A Report to the President* (Washington, D.C.: U.S. Government Printing Office, 1945).

they depended. To enter a new field, they would hire a group of new employees and invest in the necessary facilities. Ford's River Rouge plant represented the epitome of the self-sufficient approach—raw materials entered the complex at one end and finished cars left at the other.

Government R&D funding grew very rapidly during the golden age, especially after the launch of Sputnik in 1957. Soon thereafter, agencies such as NASA and the Defense Advanced Research Projects Agency were established, and funding at NSF grew very rapidly. During the late 1950s and early 1960s, federal funding of civilian R&D grew at rates nearing 35 percent per year. Not unreasonably, the attention of university researchers focused almost entirely on the federal government as a source of funds.

As the 1960s wore on, political and social turbulence began to envelop university campuses, leading to a significant estrangement between academia and the corporate sector, even though the engineering schools by and large were not much engaged in the dialogue about the Vietnam War or civil rights. Science schools were, and for them it was a very tough time.

As a result of these trends, R&D cooperation among different institutions was very limited during the golden age. It was limited because industry thought it should be self-sufficient, because government was funding the universities so generously that most faculty wanted little to do with industrial funding, and because the campuses were not congenial to an industrial presence. During this era, what industry wanted from the universities was its graduates and its published papers. Meanwhile, the government was pursuing its own objectives: hands-off basic research or highly focused mission research, largely in the field of national defense. So universities, industry, and government were proceeding along three independent tracks with relatively little interaction.

During this period, there was a limited amount of cooperative industrial R&D under rather special circumstances. For example, the chemical and petroleum industry agreed to support development of the thermophysical and chemical properties of materials at several universities. Some regulated industries also supported joint R&D in such institutions as the Institute of Gas Technology at the Illinois Institute of Technology and in the Electric Power Research Institute. A few traditional sectors whose firms had little or no indigenous R&D capability also supported joint R&D efforts, as in the textile and pulp and paper industries.

In the mid-1960s, the U.S. Department of Commerce attempted to introduce a "Civilian Industrial Technology Program" to provide federal funds to nonprofit organizations to perform R&D for lagging industries such as textiles, clothing, shoes, and home construction. This proposal was defeated in Congress, at least in part because the industries that were supposed to benefit were radically opposed to it.

I would like to mention an unusual government–university–industry contract in the mid-1960s in which the Advanced Research Project Agency supported Monsanto and Washington University to establish a Materials Research Laboratory in the field of polymeric composite materials. I served as a member of the laboratory faculty, where we often could not tell whether our colleagues were acting as Monsanto employees or as graduate students at the university, and whether some of the faculty were acting as faculty or as Monsanto employees. This ambiguity made some at the university nervous— today this would be quite ordinary, but at the time it was very unusual.

The laboratory made some important contributions to composite materials technology. More important, however, in my view is that it taught Monsanto and Washington University how to work together and enabled them to build a trusting relationship. It was key, I believe, to the very large investment that Monsanto later made in life sciences research at Washington University, which in turn was the key to enabling Monsanto to make its path-breaking transformation from an industrial chemicals firm to a life sciences-based firm.

THE POST-OIL-EMBARGO ERA OF EXPERIMENTATION

In the decade after the 1973–1974 oil embargo, the consensus reflected in the Bush report dissolved. Companies became disillusioned by the failure of fundamental research to deliver a constant stream of breakthroughs. Major corporate R&D laboratories were closed, and R&D was decentralized. Because of the economic recession of that era, federal R&D funding began to shrink, and universities began to look for other sources of research support.

In addition, the end of the Vietnam War opened the way to a rapprochement between the universities and industry, and it became possible to work together again. The energy crisis stimulated many universities to engage in research into practical industrial topics such as energy production and conservation. Paradoxically, in the 1960s universities had to learn to become "relevant," and helping with energy and environmental technology became a new focus for their societal engagement.

Furthermore, the challenge from Japan began to emerge during the end of this period, even as firms believed themselves to be under constant challenge to meet new environmental, safety, health, and energy conservation standards.

All these factors led to a new interest in cooperation in research. One bit of evidence for this change was that the Carter administration's Domestic Policy Review on Industrial Innovation conducted from 1978 through 1980 recommended that the federal government support generic applied research for industry and that the antitrust laws be modified to facilitate horizontal cooperation in R&D among competing firms.

In the late 1970s, MIT established a research center in the field of polymer processing with financial support by government and with financial support and active participation in the research by industry. This activity set the stage for much of the government support of industry–university cooperative research in the ensuing decade, including the highly successful Engineering Research Centers program of NSF.

The Bayh–Dole and Stevenson–Wydler acts of 1980 created new incentives for universities and federal laboratories to work with industry on R&D and technology transfer. Stevenson–Wydler also authorized NSF's experimental program in University–Industry Cooperative Research Centers and gave the Department of Commerce the authority (never exercised) to fund Cooperative Technology Centers to help industry with research. The National Cooperative Research Act of 1984 provided a "safe haven" for industrial research consortia to help them avoid challenge under the antitrust statutes.

William Norris, CEO of Control Data Corporation, became a convert to the cause of cooperative industrial research and was instrumental in the creation of the Microelectronics and Computer Technology Corporation in Austin, Texas, in which a number of companies conducted joint R&D related to computer technology. He based his concept heavily on his reading of the role of government-supported cooperative R&D in Japan. Similar efforts were initiated leading to SEMATECH in the semiconductor chip industry and to the National Center for Manufacturing Sciences in the discrete parts manufacturing sector. Each of these efforts began with representatives of firms in the same industry cooperating to address common problems and opportunities, with the expectation that each would commercialize the results independently. This kind of R&D cooperation among competitors is sometimes referred to as "horizontal" cooperation.

Thus, by the early 1980s, cooperation in R&D among firms, as well as between universities, industry, and federal laboratories, had begun to be viewed as quite normal and reasonable, although the number of participants was still relatively limited.

THE COMPETITIVENESS CHALLENGE

The fourth era, a period during which industrial R&D was heavily shaped by the "competitiveness crisis," overlaps to some extent with the prior period of experimentation with new forms of R&D cooperation. Many causes can be adduced for the poor performance of the U.S. economy during the early 1980s. Some blame the monetary measures taken in 1979 by the Federal Reserve Board to break the endemic inflation of the 1970s for the resulting high interest rates, high unemployment, rapidly growing negative trade balance, and loss of manufacturing jobs during the early 1980s. Others argue that the need to take strong macroeconomic action was itself a reflection of lagging U.S. industrial performance, where ineffective R&D, sluggish commercialization, low rates of investment in modernization, and slow productivity growth all contributed first to the "stagflation" of the late 1970s and then to the poor performance in the early 1980s. Still others argue that "unfair" Japanese trade practices contributed to the worsening trade balance.

One popular interpretation of the competitiveness crisis was that U.S. firms were not able to transform the results of R&D into marketable products quickly enough, especially as compared with Japan. Following this diagnosis, the prescription was to find quicker and more effective ways for industry to exploit the results of research at universities and in the federal laboratories, as well as the results from the companies' own fundamental research labs.

One response to this perception was the passage of the Federal Technology Transfer Act in 1986. This act directed the federal laboratories not only to transfer technology to industry, but also to engage in "cooperative research and development agreements" with firms and consortia. Later, the Omnibus Trade and Competitiveness Act of 1988 created the Advanced Technology Program (ATP) at the National Institute of Standards and Technology under which the federal government would help finance precompetitive R&D at firms and consortia. ATP ushered in the era of cost-shared industry–government collaboration in R&D that reached its high point with the Technology Reinvestment Project (TRP), a short-lived program to fund dual-use technology on a cost-shared basis. TRP was embraced publicly by President Clinton in March 1993 and soon thereafter became a target of partisan attack. Although the TRP only lasted about two years, it helped shape our view that government, industry, and universities should collaborate on R&D for mutual gain.

Thus, by the end of this era around 1993, cooperation had become widely accepted as important, but the dominant model remained horizontal cooperation on R&D on shared problems with common objectives.

THE ERA OF INDUSTRIAL RESTRUCTURING

We are now firmly engaged in the era of industrial restructuring, which began in earnest in 1992 or 1993. The past half dozen years have witnessed a remarkable recovery of the American economy. Productivity is up, unemployment is at historic lows, inflation is in check, the stock market is up, and interest rates are low. The information technology revolution is transforming our way of life, how we do business, and how we govern ourselves. New ways of managing businesses and new approaches to quality performance have restored American firms to their old positions of world leadership.

These new developments have had major impacts on R&D and especially on cooperation. By virtue of the determination to emphasize core competencies and outsource everything else, firms have largely abandoned the idea of self-sufficiency. Strategic alliances have become the norm, nowhere more so than in research and technology development. Firms enter into a plethora of agreements for joint R&D, not so much with competitors as with firms, universities, and government laboratories that can bring skills and capabilities to the enterprise that it would be too expensive or take too long to develop in-house. R&D collaboration has shifted from a horizontal to a "vertical" approach in which firms and others along the supply chain work together to bring new technologies to the market faster and with higher performance.

In the vertical model of R&D cooperation, universities are no longer seen as places at which to support long-range, blue-sky research but rather as essential suppliers to corporations of key research activities that are necessary to accomplish the company's main business. As suppliers, universities find themselves working under tight contracts, with deliverables and due dates, and with partners having expectations that intellectual property agreements will serve the needs of the corporation. As a result, the management of intellectual property in university–industry cooperative arrangements has become increasingly contentious. Furthermore, there is some indication that firms are beginning to focus their cooperative arrangements on a small number of universities that are viewed as institutional strategic partners. Such arrangements go beyond agreements with individual faculty members. In some research that my colleagues and I have been doing, we have found that purchasing managers are helping manage R&D cooperation— further evidence that R&D partners are being viewed as suppliers.

CLOSING OBSERVATIONS

R&D cooperation has become the norm in universities, industry, and federal laboratories. R&D leaders are deeply engaged in cooperation, and the old single-investigator model of a research career is in decline. A well-functioning technology transfer operation is now essential for universities and laboratories, and the job of external technology manager is of increasing importance in firms. Today's industrial R&D organizations follow a cooperative model that is the antithesis of Bush's model with its outmoded focus on separation of roles, self-reliance in industrial R&D, and "basic research is best." In the new environment, there has been little demand for change in federal R&D policy since 1992, other than calls for greater federal spending on R&D and for a more thoughtful allocation of existing federal funds among national security, biomedical, and all other national objectives.

DISCUSSION

Ashok Dhingra, Dupont: In regard to the Monsanto–Washington University collaboration, how would you rate that partnership as to its relevance to the commercial success of the projects that they had at that time? The Japanese took over the carbon fiber technology, which was being developed under this program. And then Dupont took the leadership for the organic fiber.

From your perspective, what were the successes of what I think was the very first lasting partnership between university and industry? How did society benefit from that long funding and high-quality research?

Christopher Hill: I assume that you are speaking specifically about the composite materials, not the biomedical collaboration. The important work in the composites area probably had to do with the development and elaboration of the field of the micromechanics of composites that was done on the mechanical engineering side of that partnership.

It was interesting because it was both chemical engineering and mechanical engineering mechanics on the university side. On the chemical side a lot of the work on naturally occurring composites, that is to say the two-phase polymer systems that John Cardose and his students did, has been important. The work that I was involved in was much more limited. I didn't make much of a contribution. We did some of the first controlled rheology of high concentrations of fiber-filled suspensions, but this was all normal science and not terribly interesting.

Hank Whalen, PQ Corporation: You mentioned that there has been no change in science policy since approximately 1992. But Vernon Ehlers, vice chair of the House Science Committee, did a study on national technology policy, the summary of which came out last September. It's my understanding that the published copy of the final report will be issued shortly. This report may not change anything, but it is at least an attempt.

Christopher Hill: Vernon Ehlers, a member of Congress from Michigan, was given the formal task by then-Speaker Gingrich to come up with a new Vannevar Bush report. I was given that charge once, and I have been part of other groups that were given that charge, and many others have tried to do it.

I argue that you cannot do a new Bush report until the world has decided that there is something new to say. Bush did not invent his new world out of whole cloth. And everyone who has said, "Let's get ten smart people around a table for six months and invent a new world of research," always fails. It is just too hard to do. So did Ehlers and his colleagues come up with a new Bush report? The answer is no.

My major point at a recent symposium on Ehlers' report at the American Association for the Advancement of Science was that they managed to make a nice statement on U.S. science policy circa 1987. Ehlers' report suggests that we could all cooperate. That would be good. But we've been cooperating since the early to mid-1980s. Second, the Ehlers report is almost completely oblivious to the emergence of the Internet. All it says about the Internet is that it's a neat way for scientists to exchange data.

3

University–Industry Strategic Alliance:
A British Perspective

P. O'Brien, William A. Wakeham, and J.T. Walsh
Imperial College of Science, Technology, and Medicine, London

The marked shift in United Kingdom and European government policies away from the direct support of research within the higher education sector has led universities to alter the focus of their search for funding much more toward industry. At the same time, governments have encouraged increased collaboration and the exploitation of the results of university research. These changes present new challenges for universities in the maintenance of academic freedom and excellence in scholarship that constitute their primary mission.

We describe how one European institution, Imperial College of Science, Technology, and Medicine, has been developing new approaches and initiatives that provide new opportunities for supporting fundamental research. Multidisciplinary strategic alliances represent an approach that has allowed the college to develop research in a commercial context while maintaining its academic purpose and excellence. However, it must be recognized that making such strategic alliances work successfully can be a difficult process requiring universities to establish new practices and methods of organization to take advantage of the industrial funding of research which will become increasingly important in the future.

INTRODUCTION

In this chapter we necessarily write from the perspective of a university with the highest research income in the United Kingdom and one with a relatively small endowment. We also write within the funding context of the United Kingdom, which has been following a particular trend that has been at some variance with the remainder of Europe, although there are signs now that other countries are following the same track. Furthermore, even within the United Kingdom, Imperial College is at one extreme of a spectrum of higher education institutions by virtue of its concentration on science, technology, and medicine and its focus on research. Thus, whereas our description of the behavior of Imperial College does not describe an average institution in a typical European country, it does indicate the direction for all European countries and institutions.

In Europe, higher education has always been funded directly by the government in large measure.

Until very recently, nearly the whole of the tertiary education system, including the staff of universities and their infrastructure, their research, and the undergraduate and postgraduate students, were all supported by central and, in some European countries, local government. Indeed, in some countries such as Germany, the academic staff members of the university are government employees. Government funding of a particular university amounted to as much as 85 percent of the university's income until the early 1980s. This funding has always been supplemented for research with grants from individual government agencies on a competitive basis, and governments of different countries have continued to support research in a collaborative manner through the European Union (EU).

In general, within Europe, the spending of gross domestic product on research is less than that spent in the United States and Japan, although there is a considerable variation within Europe and the gap is narrowing somewhat. However, one feature within the countries of the EU is a trend from high- to low-tax economies. Such a shift has made it harder for governments to justify funding the costs of student tuition and subsistence and the direct and indirect costs of research from which companies, rather than individual taxpayers, are seen to benefit. The United Kingdom, in particular, has experienced this change, and it has culminated in the introduction of student fees payable by individual students from the United Kingdom and even higher fees for overseas students. At Imperial College the transition has been such that, in place of 85 percent of the costs of the university being met by governmental direct funding, the figure is now below 40 percent, as illustrated by Figure 3.1, and has not been matched by an increase in the sums available competitively for research funding from government agencies. In some Southern European countries, this trend has been much less pronounced.

The direct consequence of this reduction is threefold depending on the nature of the university. For those universities with little research, a dramatic increase of student numbers (home or overseas) funded by fees and/or the central government can assure financial viability. Other universities with large independent endowments are cushioned against hard times by their assets and given the opportunity for further growth. For Imperial College, the relatively small endowment and the impossibility of a substantial growth in numbers of high-quality students has meant that the college has to focus its

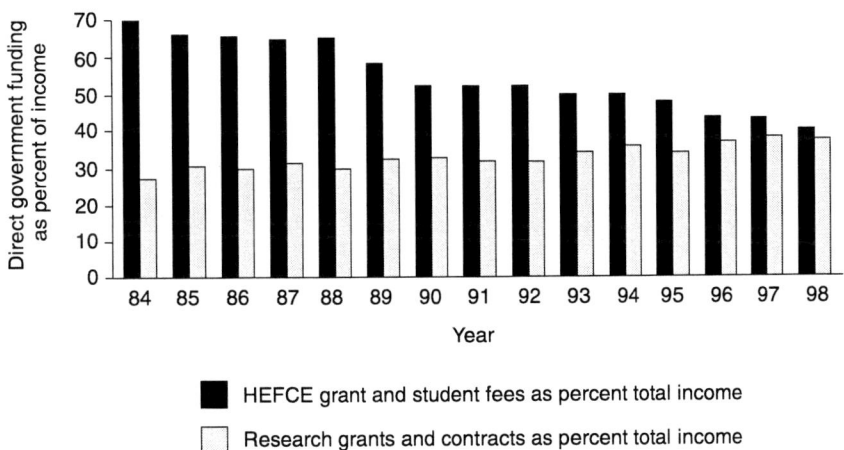

FIGURE 3.1   Percentage of Imperial College income met by government funding, 1984–1998. HEFCE, Higher Education Funding Council for England.

reactions to these developments. To ensure its development it has relied on its research base to finance its activities through new mechanisms. These new mechanisms have included the search for new opportunities and new markets for science and technology. The most significant example of this is Imperial's creation, by merger, of the largest medical school in Europe. This has, in part, been driven by the belief that medicine can benefit substantially from the interaction with basic science and engineering and also by the pragmatic observation that future government funding will increasingly be in health and related activities in the new century.

A NEW CLIMATE

The changes described above affected the composition of research income and altered the proportion of funds from various sources, so that today, for Imperial College, 60 percent of research income is derived from non-U.K. government sources, as shown in Figure 3.2. This has in turn ushered in other more far-reaching developments. Hitherto its reliance on central government funding induced a culture of dependency and militated against risk. Universities were conservative institutions intent on protection of intellectual property because any alternative was thought to be a potential misuse of funds. In the new climate, the diversification of our funding base has meant that the college can now take advantage of new opportunities and is no longer prevented from taking (albeit calculated) risks.

The environment in which universities operate has also been modified as a result of other government policies. Attempts to increase the fraction of the population engaged in higher education in the United Kingdom from 25 percent to 40 percent has shifted the focus from research-based universities to

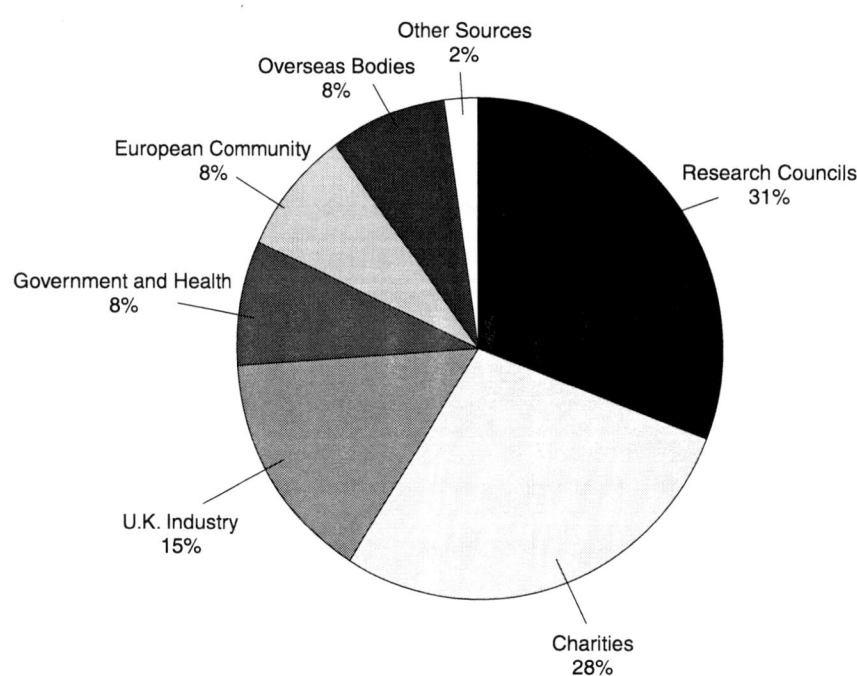

FIGURE 3.2   Research income for Imperial College, 1998.

teaching-based ones. Although the financial provisions made for these additional numbers ensure that teaching-based universities are financially viable, the system now makes it increasingly difficult to have a highly focused teaching university that is also active in research. Government initiatives to support research have changed the conditions even further, emphasizing the importance of industry as both a stakeholder in higher education and as a potential source of funding. This is reflected in the initiatives that have been forthcoming.

GOVERNMENT INITIATIVES

One recent initiative has recognized the relative lack of a venture capital market within the United Kingdom and thus the government has organized a competition for seed capital funds for universities. The successful universities have been granted a capital fund, which is simply constrained to developing the intellectual property they own. The mechanisms can range from the development of patented ideas toward the marketplace through licensing, to the creation of spin-out companies. This concept moves as far as possible from the tradition of risk avoidance. Other, longer-term programs have also been developed to support the government's increasing expectation that the results of research that they, or others, sponsor should be exploited and marketed. As well as this stimulation of an increased exploitation of research results at the output stage of the research process, governments have become increasingly keen to stimulate interaction between industry and universities at the input stage that will allow research ideas to develop and be exploited in an industrial context. One example of an intergovernmental approach of this kind is provided by the Fifth Framework Programme of the EU. This is the latest of a series of programs of technical endeavor funded by the EU but it is a far cry from the early programs whose main purpose was the stimulation of international contacts in science. It is an essential part of the latest program that there be industrial involvement and an exploitation strategy.

Within the United Kingdom, the so-called "Link Programme" is another example of a government initiative that brings industry and universities into more intimate contact. It functions through a consortium of industrial companies funding precompetitive research jointly with the government. This sort of funding is particularly associated with the Foresight Programme organized by the previous United Kingdom administration. In this program all interested parties including industry, government, and universities were invited to contribute through various sectoral panels to identify the top priority areas for research in the country. This process allows targeted funding for these research areas without provision of new money. Since 1995 this has lead to about 20 very innovative research partnerships between industry and universities, each with a value of roughly £200 million, which is equivalent to two-thirds of the annual United Kingdom budget for physics.

"Realizing Our Potential Awards" is a further illustration of research support that, in this case, seeks to encourage universities to undertake research with a relationship to industrial needs. Specifically this scheme offers a "blue-skies" research grant for an investigator who has held a qualifying applied or strategic industrial research contract funded by industry. The work carried out under such a grant is itself not permitted to be applicable directly to industry, but there is obviously a strong indirect incentive encouraging such work.

INFRASTRUCTURE FOR RESEARCH AND INDUSTRY

The frequency and range of these initiatives, as well as their specific nature, has encouraged universities to adopt quite different structures than have been traditional to manage their operations with regard to research and industry. To enable us to participate in these government initiatives and to take maximum advantage of industrial opportunities, Imperial College has developed a support infrastructure for managing research contracts and grants. Thus, on top of the usual activities associated with the negotiation and agreement of many kinds of contracts for research with a variety of government and nongovernment agencies, the college has added specific entities concerned with intellectual property exploitation, the provision of scientific services, and consultancy. It is the view of the college that it is no longer acceptable for these activities to be conducted in a rather amateur fashion but rather that they must be pursued with full commercial rigor. For this purpose it is necessary to employ staff with industrial training and appropriate professional qualifications. Necessarily, these staff members have a different culture from that of the traditional employees within a university and there are tensions created that have to be managed, not least with respect to remuneration.

An important element of this increasingly professional approach to research management has been the desire to evaluate the full costs of all activities, including research. Imperial College has been at the forefront of this activity-costing exercise in the United Kingdom and has aggressively pursued policies of the full funding of research by all sponsors. Combined with a rigorous policy of intellectual property ownership and high quality of research, this approach to overhead recovery has more than maintained the financial situation of the college and developed intellectual assets of considerable value.

For some considerable time the college perhaps cared rather more about the protection of those intellectual assets than about their exploitation, partly because of the aversion to risk. However, this view has been moderated recently as pressure from the academic community and government to see the results of research exploited has increased and the reasons to fear risk have diminished. Instead the college now takes the view that the exploitation of intellectual property generated by staff is an important factor in retaining high-quality staff. Indeed, this is seen as the primary motive to seek exploitation. Whereas there is always the chance of generating income from licensing or equity growth in companies from exploitation, the former route is safe but seldom lucrative and the latter has a very small chance of success. Indeed, it is well known that rather few technology transfer offices of universities cover their costs, as the chances of great success are small. Neither of these is, in the end, a primary motive for the college to seek exploitation of its intellectual property rights, although, of course, the college would not spurn such success or income!

Imperial College has a technology transfer company (IC Innovations) that is staffed with technology transfer professionals who identify the commercial potential of research and market it, sometimes, when appropriate, by forming spin-out companies or, more often, by licensing. When a spin-out company is created, the college normally takes a share in the equity, which is justified by its investment of its intellectual property rights in the spin-out. However, the share is usually, but not always, a minority one, and the independent companies are left to exploit the intellectual property rights in the best commercial manner possible with a strictly arm's length relationship with the college. The college has a wholly owned subsidiary that can provide its spin-out companies with legal and business mentoring services and advice; it also aids them in their search for venture capital.

In addition, the college operates a consulting company called IC Consultants (ICON), which markets the expertise of the academic staff as consultants. It accepts the professional liability for the provision of such consulting activity and negotiates the fees for the work of individuals and groups. ICON also markets the use of the college's scientific research facilities for use by industrial partners when they are not in use for their prime purpose of college research. Thus, for example, unique wind tunnel facilities have been made available at commercial rates for the development work of Formula-1 racing teams in a manner that provides funding for future college research in a complementary fashion. This particular college company has been important in establishing linkages with small companies with fewer than 200 employees. Such companies have proved to be rather difficult for Imperial College to reach and engage in research. In part this is because, for small companies, the typical size of a research project is a large sum of money but also because they are interested in focused short-term work rather than longer-term strategic research. ICON is a further example of the college's commitment to see the ideas and expertise of its staff members exploited for their own benefit and that of the college itself.

STRATEGIC ALLIANCE

An even more recent addition to the armory of means of interaction with industry is the rather extensive mode of partnership between a company and a university that we term a "strategic alliance." The concept of the strategic alliance is a relatively new concept in Europe in terms of the relationship between universities and industrial partners. The characteristics of such an alliance are that it will be a long-term partnership across a number of the university departments or disciplines, usually beyond the life of a single research project contract, and will normally involve a large multinational company. The alliances are open ended with no limitations on what sort of interaction might be involved so that it can include staff exchange, undergraduate recruitment, and student prizes or endowments. An open negotiation is fundamental in getting an alliance to work and must proceed from a position of mutual trust toward an agreement for mutual benefit. Such mutually beneficial arrangements do not follow automatically from the signature on an agreement. Rather, they take a considerable amount of work on both sides to stimulate and refresh the interaction. When they work, the benefits are substantial to both sides because the process builds a relationship that has a broad base within both the industrial organization and the higher education institution and a degree of interdependence founded on respect for what each party brings to the relationship, which makes it more than the relationship of a provider and a customer. For example, universities generate many of the new ideas and the future employees needed to perpetuate a vibrant industrial sector, but it is in the interests of both parties to ensure that graduates are employable and that ideas with potential are commercially exploited.

One example of such a strategic alliance at Imperial has been the involvement of two pharmaceutical companies, Smithkline Beecham and Zeneca, who have supported not merely a research project, but the entire discipline of analytical chemistry by funding a center at Imperial in that topic. This was because the two companies believed that the national need for people with such skills (which was particularly acute, although not confined to them) was not being fulfilled by existing activities. The center has grown with their initial support into a dynamic center training students and staff and performing innovative research for those two companies and others. Another example, particularly in a European context, is the increasing desire of companies to have graduates with a multinational and multilingual background—something that the college also encourages. Our industrial links with several multinational companies across disciplines and national boundaries allow our students to work outside of the United Kingdom on internships or projects benefiting themselves and their future employers, whether they be the project sponsors or not. It might also be said here that the development of strategic alliances between universities in several countries with some of the same purposes in mind cannot be far away.

For strategic alliances to be successful, hard work must be added to a carefully constructed framework. Our experience has shown that commitment from a company and the university partner must be long-term, flexible, and at a high level in both institutions, and not be confined to a single project, a single academic investigator, or a particular company division or academic department. The alliance should begin with a network of collaborative projects and relationships between the two partners; and when this base has been established, further activities, funding opportunities, and industrial partners can be developed with great ease.

An essential part of the framework of the alliance has been a practical realization by the university

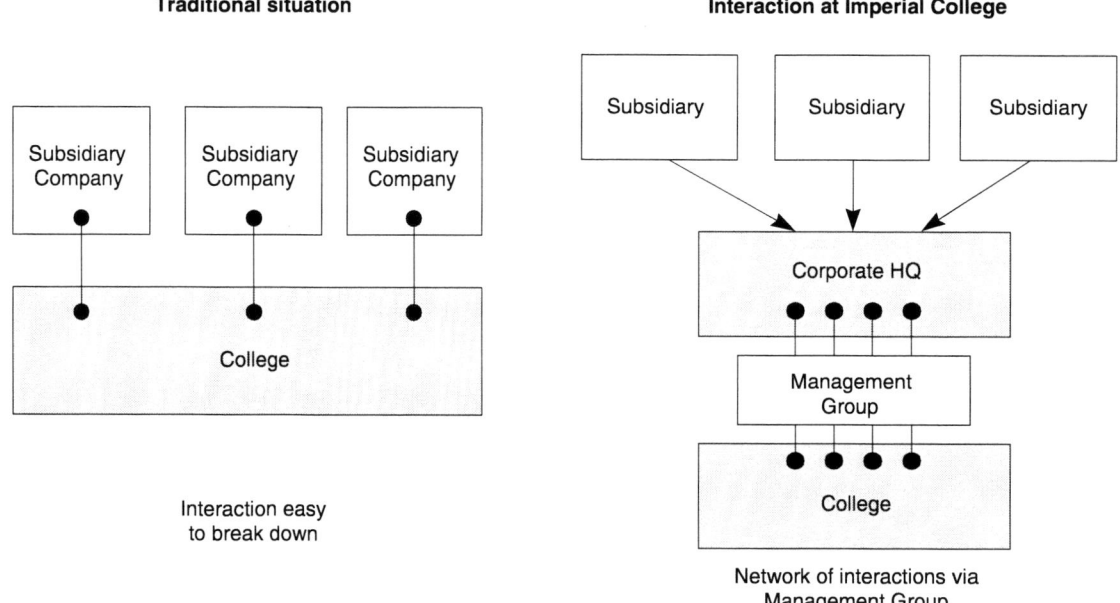

FIGURE 3.3   Strategic alliances.

of the industrial need to seek solutions without regard to discipline. This has encouraged the development of a different university structure for research from that which remains appropriate for the education of undergraduates where a single traditional discipline remains the norm. The use of a broadly based management group for each strategic alliance has also been a key part of this development. This group includes representatives from both sides involved: members of several college departments and several members of the company at the corporate level. This body has operated to remove barriers to communication between the two partners and internal barriers within the two individual components of the partnership. The inclusion of high-level representation in the group has the added benefit of a communication between entire organizations rather than individuals, which enhances the networking, the trust, robustness, and effectiveness. The change in the nature of the interactions between a university and a company introduced by a strategic alliance is illustrated in Figure 3.3, where it is compared with the more conventional interaction.

It is evident that the network of relationships between a university and a company, exemplified by Figure 3.3, has enormous benefits to both partners in the context of government support for interaction as discussed above. The alliance provides ready-made partners for collaborative ventures in which the sums of money invested by the company in the university enable other funds to come from the government and, in the European context, intergovernmental funding that makes use of the multinational nature of the industrial partner.

AN EXAMPLE OF AN ALLIANCE

Among the strongest alliances that Imperial College has established to date is that with Air Products and Chemicals, Inc., a company operating in a number of sectors throughout the world. This is a prime example of how an alliance has brought about changes within Imperial College that had been more difficult to accomplish without the interaction. The relationship has been a productive one in terms of scientific output, some with commercial potential, some effective student exchanges, and the acquisition of new research funding inaccessible without the relationship. In this relationship we jointly created an arrangement in which Air Products provides an annual sum to fund research. This sum is available to bids from academic groups within the college and judged by a panel made up of staff from the college and members of the company. For projects to be successful, they must have the support of a business sector of the company as well as a corporate drive. That support may be expressed in writing or through separate funds supplementing the corporate funding. The result of such an arrangement is of obvious academic benefit to the college and potential commercial benefit for the company. The projects have brought together members of the Chemistry, Chemical Engineering, and Materials departments at Imperial College in a new manner to achieve a common end. This particular alliance has also had a profound effect on the Chemistry Department at Imperial College since it contributed to a new-found confidence for interactions with industry. The department has raised its research income from industry to nearly 70 percent of its total research income and to a figure three times greater than five years before. Obviously, strategic alliances also have considerable financial benefits in addition to those of an academic kind.

In addition to the direct benefits, it is worth pointing out some less obvious gains from strategic alliances that are, nonetheless, important. Agreements over intellectual property and overheads can be negotiated once instead of for each individual research project; as well as being more convenient, this also in some small way reduces the college overhead itself by reducing the amount of staff time spent on negotiation.

ACADEMIC FREEDOM?

One of the most obvious concerns when pursuing any kind of collaboration with industry is that there will be a divergence of motives that could lead to a loss of academic freedom. However, as illustrated above, the basis for a strategic alliance is one of mutual benefit, only where a convergence of interest between the college and the company is feasible. In addition to mutual interest, cooperation and trust are equally important in the foundation of such an alliance. These elements of the agreement are usually enshrined within an overall memorandum of understanding that spells out the rights and obligations of each party. Although such agreements will usually encompass general conditions surrounding the alliance, they will not include any restrictions on the college in terms of other collaborations or areas of activity. Such conditions on confidentiality as there may be will be held within individual project agreements. Thus, there is no compromise of academic freedom and, indeed, it may be argued that such freedom is enhanced by the provision of additional income, especially that which meets overhead costs. Furthermore, because some government schemes reward industry-supported work with a blue-skies grant, there is a further stimulus to independent, innovative work.

CONCLUSIONS

The reduction of centralized governmental funding for universities within Europe, and the United Kingdom in particular, has altered the relationship between industry and higher education institutions. It has proved possible for a research-led university to develop a relationship with several multinational companies that enhances its research activity and enhances the relevance of its research without any compromise of its scientific freedom. These relationships have several characteristics in common, which have encouraged multidisciplinary approaches to research and the development of long-term interactions with a greater stability than achieved by isolated project grants. Indeed, government measures within the United Kingdom and the EU have introduced new schemes that make government funding more easily secured with such alliances in place.

DISCUSSION

Fritz Kokesh, Massachusetts Institute of Technology: You indicate that over a period of time, Imperial College's position on intellectual property has changed. Could you explain this further?

William Wakeham: Essentially, if we have to pay a fraction of the cost of the research, meaning if the industrial company does not pay entirely, we would expect to own the intellectual property. The more that industry pays toward the full costs, the more they can share in the intellectual property. If they pay the full costs, there is a reasonable argument that they may be allowed to keep it as long as they promise to exploit it.

One of our concerns is that many companies choose to keep the intellectual property in order to not exploit it, to prevent somebody else from exploiting it. So we usually try to write provisions into the agreements that someone must exploit it.

Fritz Kokesh: Would you explain more fully about what full cost means?

William Wakeham: We have worked out our full overhead rate using fairly simple, but government-approved, rules as to how to cost our activities. We know, for example, by survey, how much time our academic faculty spend on research and how much time they spend teaching. Once you have this information, you can cost all of your research activity and prove what the costs are. And in negotiations with a multinational company, you have to prove this and a lot of other things about what you say. If you say your overhead is "x," they want you to prove it. And you have to be able to. Traditionally this is not easy for a university. One of the points I made in my presentation was that we employ people who are from industry in our industrial liaison office. They can do these calculations in a hard-nosed way and in a way that the people on the other side of the table will appreciate.

David Schetter, University of California, Irvine: Are the faculty allowed to strike their own independent consulting agreements under this icon, or is it a requirement of the institution that everyone participate? Second, how do you initiate a strategic alliance? Does identifying opportunities come from the faculty up, or from management down? How are those put together from the start?

William Wakeham: The consulting company is not obligatory. Perhaps I should say that, in a British context, the academic people are much less aggressive in negotiating their fee. They are prone to do things out of interest, not to make the right amount of money from it.

One of the uses of an icon is to negotiate the fees. It also negotiates a cover for professional liability and provides a separation between the university and the liability. In a test case at one of our universities, an individual member of the faculty did a consulting job, got the wrong answer, was sued by the company for getting the wrong answer, and the university was found liable even though the faculty member never used the university note paper, telephone, or anything. This is a dangerous precedent. So we now use a company to separate the university. The consulting is taken by the company and subcontracted to the individual. And, of course, if you are an individual member of the faculty, you can see the advantage in this arrangement.

Regarding your second question on who initiates strategic alliances, it can be either from the faculty up or management down. In the case of Air Products, it took place because one of our alumni was a vice president of research at Air Products and suggested that it would be a good idea to get back together at Imperial College. But in some cases, it's been a top-down decision by someone such as myself, saying that we should talk to Smithkline Beecham about analytical chemistry, which in our case disappeared into almost nothing inside the Chemistry Department. We wanted to build up again. It turned out that the department could see a national need for that because no one was being trained in analytical chemistry. Several pharmaceutical companies then came together and agreed with us. It wasn't sponsoring particular research projects, it was the existence of the discipline that was supported. So there has to be a synergy between what the companies want and what the college wants.

Janet Osteryoung, National Science Foundation: Have you run into any problems with some of your external partners bumping into each other inside the university?

William Wakeham: Yes. Our linkage with British Petroleum (BP), for example, which is in the area of catalysis, essentially supported about 30 people in one laboratory in a particular area. The type of catalyst being studied was of interest to more than one company. Imperial Chemical Industries (ICI) was also interested in that catalyst, but for a polymer that was of no interest to BP. So we tried very hard to get the companies to understand that they could work together. In the end it failed. ICI would not be involved because BP was already involved, even though the technologies coming out of the alliance would have been completely independent.

Panel Discussion

Laren Tolbert, Georgia Institute of Technology: One thing that was not mentioned about the so-called "home-run" patents, which were largely biomedical, and more specifically pharmaceutical, is that these areas produce a single compound. This has a very clear patent situation with a clearly identified profit margin, whereas many other patents represent an improvement on a process. The same may be said of software, although that is still an emerging field.

Should we separate the way we treat patents in these different fields to avoid a one-size-fits-all approach to intellectual property issues by making every innovation fit the biomedical model?

David Mowery, University of California, Berkeley: Public universities, and public institutions in general, often find themselves in a more complicated situation because the political overseers will tend to insist on accountability of various forms, which pushes people toward a uniform policy. So there will always be attention there for public universities. But by and large, it may make a lot of sense to stop looking at the entire world as though it's biotechnical or biomedical.

Christopher Hill, George Mason University: On top of that, software can be either patented or copyrighted or both, and the presumptive rights to copyright ownership can be different from the presumptive rights of ownership of the patent. Furthermore, it appears now that undergraduate students who work on software as part of a homework assignment that is later integrated into a software package may have a copyright interest. We are just beginning to thrash this out. So I don't think that anyone pretends that one size is going to fit all.

William Wakeham, Imperial College: The point about students is quite interesting. Upon registration, we require a student to assign any of their intellectual property to the college. This is not necessarily legally possible, but we do it anyway.

But there is a very big difference with the biomedical area because the patent situation is usually very different than in, say, mechanical engineering, where a particular device is created. This usually stands alone. However, in the biomedical area, the central patent has to be "decorated" with other things to build a broad enough portfolio to be useful. We have some examples of instant successes, but mostly it has to be built up slowly. There is a lot more work in developing something in the biomedical area, but also there is a much bigger profit if one gets all the way to a patent. With software, once the code is written and professionalized, it's all ready to go.

David Schetter, University of California, Irvine: Dr. Mowery, is it not the case that, prior to the Bayh–Dole Act in 1980, the universities were not allowed to own the results of federally funded research, and that a primary impetus of Bayh–Dole was to create an incentive for universities to commercialize federally funded research? Second, if that is the basis of Bayh–Dole, do you have any data to determine whether or not it has achieved its goal?

David Mowery: Universities were, in fact, able to obtain patents on and license the results of federally funded research before Bayh–Dole under the terms of what were known as institutional patent agreements, which were negotiated by individual universities with individual federal funders of research. In fact, one of the pressures building up in the system before Bayh–Dole, notably in the University of California, was from what was then Health, Education and Welfare, overseer of the National Institutes of Health (NIH), to prevent the universities from negotiating exclusive licenses for the results of NIH-funded research. So universities were already in the licensing game selectively, although not nearly as many and under much more complicated and perhaps less stable rules than what came about in 1980. There can be little doubt that Bayh–Dole has brought more universities into technology transfer and licensing, especially when you look at the number of universities that have established offices and the amount of patenting that has been conducted. Do we know that patenting of university-developed technology resulted in either a more rapid or a successful transition to commercialization? This is a much more difficult question to answer. And it is a question that will almost certainly vary among universities, among technologies, and a host of other things. It's very difficult to measure because it is an experiment without a well-developed control group.

Christopher Hill: Another impetus for Bayh–Dole was the Department of Energy (DOE) Act of 1977, which had an explicit provision that was exactly contrary to Bayh–Dole that prohibited ownership of patents by any contractor or grantee. And because DOE was then so important, it had undercut the system you are talking about in a serious way.

Cheryl Fragiadakis, Lawrence Berkeley National Laboratory: One big contrast between the U.S. system and its history and what we just heard about what's happening at Imperial College in the United Kingdom is the rather xenophobic approach that the U.S. government seems to overlay on research collaborations and licensing versus the very forthright desirability of multinational interactions in Europe. I see this as a limit to what is happening in both national laboratories and to some campuses that actually go by the law in the United States. Is the contrast as stark to you? Do you have any forecasts of what should be done or might happen in the future on this?

Christopher Hill: I did not hear anything in Professor Wakeham's description of Imperial College that we do not do now at George Mason, a public university. The difference is that in Professor Wakeham's presentation, "multinational corporation" was a euphemism for an American company. But we would have no problem doing a strategic alliance with an overseas corporation at this point. If we wanted to accept federal government money as part of the package, then we would have to fulfill some of the requirements in U.S. law written by Patrick Windham to deal with the xenophobic impulse.

Michael Marron, Office of Naval Research: This question is directed to Christopher Hill. You ended on a note of increased private investment in research as a rosy picture but you gave a negative picture of the role of government involved in these areas. When those of us in the government enterprise look at the studies done by the National Science Foundation (NSF) on how much the investment has grown in various sectors of the economy, we're taken aback by the inability to assess how much of that research investment is highly directed. And I'm trying to avoid the applied research versus basic research polarization, but rather highly directed research as opposed to the undirected broader base research which we in the Defense Department (DOD) refer to as 6.1 and 6.2 type of research for our technical base investment. We are faced with people saying, "Look at all this huge growth here. We don't really need you guys anymore." And we see a different story when we examine it on a case-by-case, industry-by-industry basis. I don't know whether it's appropriate, but I would like to hear your comments about how this affects this rosy picture that you see developing in the 1990s.

Christopher Hill: It is difficult to parse out what any of us mean when we talk about dollar expenditures for R&D. The official statistics on federal funding for R&D include 6.3, 6.4, 6.5, 6.6, and 6.7. However, a great deal of activity in 6.4 and 6.5 is not what universities would recognize as R&D. Four years ago the National Academies of Science and Engineering issued a report called *Allocating Federal Funds for Science and Technology* that advocated the use of a new construct that it called federal science and technology (FS&T), which is an attempt to correct the statistics to eliminate those things that go beyond R&D, such as testing and evaluation of new systems. By that measure, FS&T funding, instead of being $75 billion, is around $45 billion, and it has grown more slowly than R&D. On the industrial side, the same is true. The largest funders of industrial R&D are the car companies: General Motors, Ford, and Daimler-Chrysler. My guess is that a great proportion of General Motors' $5.5 billion or $6 billion R&D budget looks like DOD's 6.4 or 6.5 activity. It includes testing prototypes of new vehicles, which is not unlike what the universities would say is not R&D at DOD. It is not really "R&D" at General Motors either. But it is for tax purposes, and so it is reported as such to NSF. So we have to be careful about statistics for both industry and government R&D spending. Nevertheless, no matter how you measure it, the federal R&D investment during the 1990s has been flat, while the private investment has grown very rapidly.

Michael Marron: My point and your point—you seem to be agreeing with me at this stage—is that heavy growth is really in the highly directed arena and not so much in the broader seed region that we might otherwise argue. And so maybe the picture is a little bit askew here when we look at the role of the federal government in this investment partnership.

Christopher Hill: I am talking about the decline of the large, centrally funded industrial fundamental research labs. Many of them are gone.

David Mowery: You are right in the sense that, if you break this down into the discredited categories, industrially funded R&D during the 1990s is by far the fastest growing category, with more than 10 percent annual rates of increase since development, which is much more rapid than basic research, whatever that may be.

Christopher Hill: I would add that, in the information technology sector, a great deal of what generates new technology is not even counted by the firms who are doing this as R&D. In Northern Virginia, there is a very dynamic, 1,500-firm information technology sector. Almost no one claims to be doing R&D, and yet it is a rapidly developing, changing industrial economy.

Fritz Kokesh, Massachusetts Institute of Technology: "Directed research" is an interesting term. To me it implies that there is a purpose in mind if it's successful. But the comments here assume that there is another feature to it that assumes that it's not frontier research. And the two aspects can be separate. To illustrate, if someone can see a purpose for something that's still occurring at the frontiers of science, then that should be considered every bit as basic as something that can't be seen.

Christopher Hill: I agree with you on that.

Thomas Manuel, Council for Chemical Research, Inc.: It's important to iterate some of the things that are implicit in some of the talks here that would refer to what I call a "social dimension" of collaboration. And the first one is that if you look at the agenda of this workshop, it's sliced into bilateral pieces. The fact is that most collaboration and the most fruitful and the greatest trend for the future is going to be tri- if not multilateral. This is implicit in Professor Wakeham's presentation.

Second, collaboration is a contact sport, and it's an iterative activity. So not only does one need patience in a particular relationship, one needs to try again and again and practice it. The third thing intersects directly with Dr. Hill's observation that everything fell apart in the late 1970s. The Council for Chemical Research attempts to continually address these types of needs and opportunities as it goes forward in many directions and as the picture changes. So there is a need for all of us to continue to work together in this area, seek ways to find new collaborations, and to not get fixed into any particular paradigm of the moment.

4

Research Teams at Universities: The Center for Interfacial Engineering

D. Fennell Evans and Matthew V. Tirrell
University of Minnesota

The Center for Interfacial Engineering (CIE) was funded by the National Science Foundation (NSF) as an Engineering Research Center (ERC) in October 1988. There are currently 26 ERCs in the United States, each of which is funded at an annual level of between $1.5 million and $3.0 million for up to 11 years. Most of these centers focus on a specific technology (for example, electro-optics, deep-sea platforms, plasma processing, and so forth). The CIE is somewhat different; our goal was to develop a new cross-disciplinary engineering field: interfacial engineering.

Competition to obtain an NSF ERC grant is extremely fierce. In the year we applied, there were 68 proposals, only 4 of which were successful. Our strategy was to suggest that establishing a new engineering field focused on interfaces addressed a pressing need for U.S. competitiveness. Our commitment was that industry would be a full-fledged partner in this endeavor. Our proposal drew on existing strengths in the University of Minnesota's Engineering and Science departments, included letters containing substantial financial commitments from 20 U.S. companies, and contained a major financial matching commitment from the University of Minnesota.

After 18 months and several site visits, we were finally funded, but only at two-thirds of the proposed level. Our first difficult task was to narrow the scope of the center's activities to match our resources. We decreased by half the number of faculty involved in the center and redlined a number of proposed programs. There is no activity in a university that is so virtuous that it should be undertaken in the absence of the required resources!

The original vision for CIE, which remains unchanged, is to

1. Lead the development of a fundamental understanding of interfacial processes,
2. Establish synergistic transfer of technology between CIE and industrial partners, and
3. Produce well-educated graduates who can apply their knowledge of interfacial operations to new processes in a productive manner.

The expectation is that the research program will be first rate, addressing issues that will impact national competitiveness and will be carried out in direct collaboration with industrial researchers. The major driving force behind the technology transfer program is that it is proactive, managed by an individual with industrial experience, located in the center, and not managed as a adjunct activity by lawyers hidden away in one of the university's many bureaucracies. The expectation of the education program is that it goes beyond the traditional university programs by introducing students to the systems approach to projects. More recently, the production of exemplary instructional materials has been incorporated into the NSF's ERC expectations.

All ERCs are exhaustively and seemingly continuously evaluated. The NSF performs an annual site visit with five to seven external reviewers, who typically produce a 7- to 14-page review. Critical reviews for continuance occur during years three and six. Several ERCs have been terminated as a result of an unsatisfactory third-year review.

In addition, the agreement with the NSF calls for an annual report containing substantial amounts of detailed data that consumes the equivalent of one full-time staff person per year to compile. Because an ERC's operation involves handling millions of dollars from the federal government, the University of Minnesota, and member companies, the activity attracts auditors like vultures to prey. Dealing with the expectations of companies and simultaneously with the rules and regulations of two major bureaucratic organizations requires a sophisticated and highly skilled staff. The costs of running an ERC properly are not trivial.

ORGANIZATION AND OPERATION

As shown in the organizational chart in Figure 4.1, CIE has four research programs, each under the leadership of a professor. Associated with each of these programs is a technical advisory committee that meets twice a year with faculty to review progress and suggest new research challenges and directions. All research projects at the CIE are nonproprietary in nature.

A major focus in the center has been the development of a Characterization Facility that currently contains over $7 million of high-level research instrumentation. This facility has become a major research resource for the university and was used last year by 400 university and faculty personnel and by 120 companies. The annual operating budget of the facility is $700,000 and is obtained from instrument-user fees. These funds cover the salaries of the facility personnel and service contracts on the equipment.

As part of establishing a new discipline, the center sponsored the publication of six textbooks in a series entitled "*Advances in Interfacial Engineering*."[1] More than 10,000 copies of these books have been sold. In addition, the center has developed several series of computer-based instructional modules in interfacial engineering, fluid mechanics, and thermodynamics. The university has licensed these to a private company for distribution.

The magnitude of the operation is perhaps best depicted in terms of the statistics applicable to our recently completed tenth year of operation (see Box 4.1).

UNIVERSITY–INDUSTRY PARTNERSHIP

There are two main membership categories:

- Sponsors, who pay $80,000 per year and are involved in all of the research programs, and

---

[1] *Advances in Interfacial Engineering* (New York: John Wiley, Inc., 1992–1999).

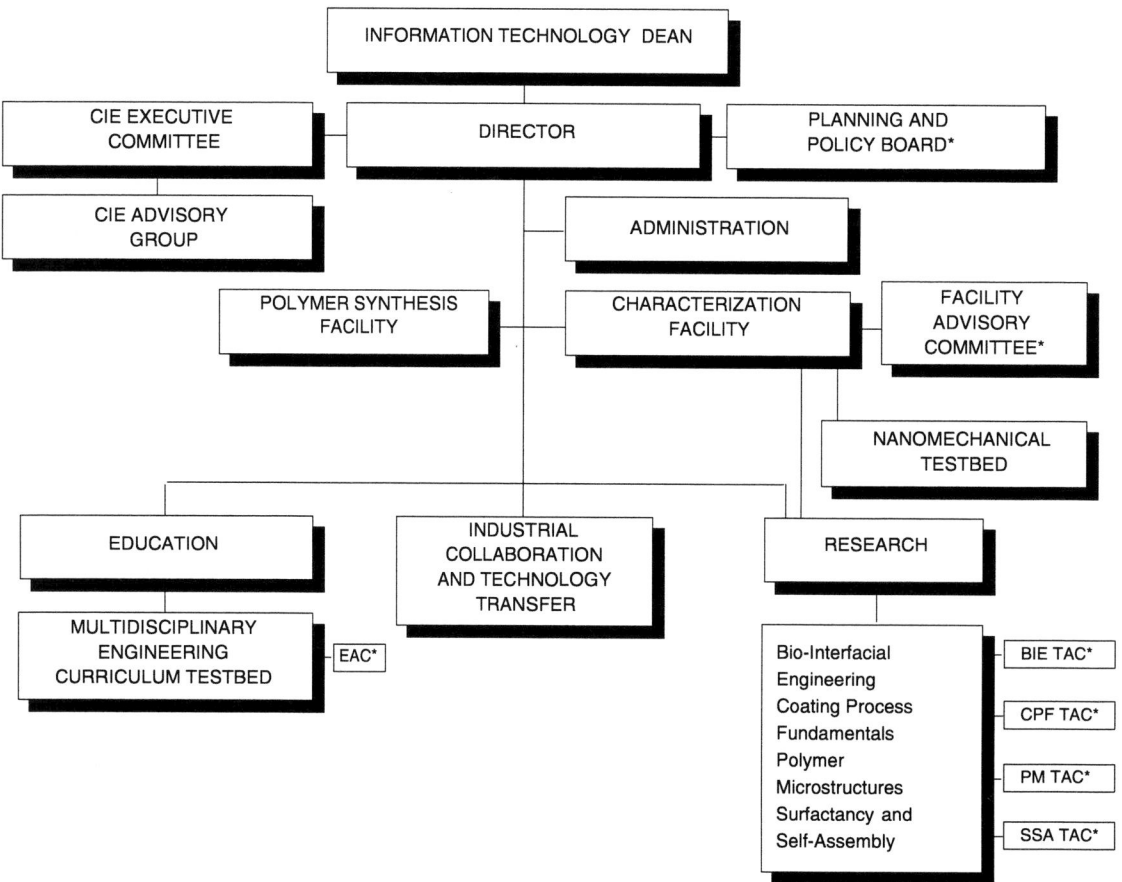

FIGURE 4.1   CIE organizations. EAC, Educational Advisory Committee; TAC, Technical Advisory Committee. Asterisk denotes industry membership.

- Affiliates, who pay 0.03 percent of annual sales with a cap of $30,000 and are involved in only one of the research programs.

Company involvement in the center is governed by a standard membership agreement between the university and the company. Obtaining an acceptable agreement is a not trivial task. We spent our first 18 months negotiating an agreement with member companies. Membership dues are paid into the University of Minnesota Foundation free of indirect costs.

According to the membership agreement, sponsors can license all intellectual property developed in the center. Affiliates are limited to developments in the program of which they are members.

A major activity in the CIE's technology transfer program is the Industrial Fellows Program. The origin of this program was the realization that many forms of technology transfer (i.e., brief one-day meetings, annual reports, reprints, etc.) are not very effective. The obsession on licensing of patents is often counterproductive. With few exceptions, most universities spend more on legal fees associated with patenting and licensing than they take in from the process. To allow effective technology transfer,

---

BOX 4.1
Statistics for Operation of the Center for Interfacial Engineering, Year 10

BUDGET

Resources:	$11.7 million
National Science Foundation: 20%	
Industry: 23%	
University: 14%	
Faculty (principal investigator) grants: 43%	
Direct Expenses:	$6.4 million
Research: 64%	
Facility equipment: 20%	
Education: 6%	
Technology transfer: 5%	
Management: 5%	

CIE PERSONNEL:		550
Faculty:		79
Industrial researchers:		84
Industrial fellows:	12	
Collaborators:	72	
Graduate students and postdoctoral fellows:		255
Undergraduates:		114
Staff (full-time equivalent):		18
Technical:	12	
Administrative:	6	

---

sponsors can send an employee as an Industrial Fellow to the center for three months or longer and receive a $50,000 rebate on their dues. These fellows work on a generic research project that has been negotiated between the center and the company. More than 140 Industrial Fellows have been involved over the past ten years. Most of the fellows continue to remain involved in center activities after their residency is concluded. In fact, there is even an Industrial Fellows alumni association. These Industrial Fellows have operated as effective change agents in our university environment. In addition, whenever they change employers they become very effective recruiters by having their new company join CIE.

Getting all of the legal agreements in place is an enabling step, not an end in itself. Our premise is that companies offer the university every bit as much as the university offers companies. Translating this belief into realization requires driving a cultural change that, if successful, leads to the establishment of a true partnership between university and company researchers. To be blunt about it, part of the challenge is to get the faculty to stop lecturing to the company researchers (i.e., to shut up and listen). The other part is to get the industrial people to talk. That seems to take about three years.

Developing strategies to sustain the dialogue between companies and the center constitutes a second major challenge. Over the years, we have developed a joint skill of translating industrial-based concerns into high-quality Ph.D. theses. The technology transfer operation plays a vital role in this activity and becomes the formal exchange point sustaining continual, more informal exchanges between researchers. We keep a log of important industrial conversations and action items; our log now contains more than 14,000 entries.

From the company perspective, it is essential that all action items that come to the attention of the center are followed up and acted upon. From the perspective of the center, it is important that commitments to companies are implemented. Having a Technology Transfer Office with extensive industrial experience and whose native language is "industrialese" is essential. We professors speak "professorese," and a translator is a necessary ingredient to building an effective and sustainable partnership with companies.

One of the hallmarks of an ERC operation is the development of a strategic plan to guide the development of the center's program and provide benchmarks against which to measure progress. A number of the achievements are summarized as follows:

- Accelerated the development of interfacial engineering as a new interdisciplinary field;
- Graduated 522 students with center experience;
- Established the Characterization Facility via a $7 million equipment investment;
- Delivered short courses and workshops to companies;
- Hosted some 123 company researchers on campus as Industrial Fellows; and
- Developed textbooks, multimedia instructional materials, courses, and practical experiences for students.

MAINTAINING THE INDUSTRIAL–UNIVERSITY PARTNERSHIP AFTER ENGINEERING RESEARCH CENTER FUNDING

At the beginning of the center's operation, we set into motion a strategic plan for sustaining the center after the end of ERC funding which, at that point, was 11 years away. This initially involved careful management of cash flows and interest on the University of Minnesota Foundation funds to generate a substantial reserve fund. More recently it has involved careful planning with the center's Planning and Policy Board as it becomes the main stakeholder in the center's operation.

DISCUSSION

David Schetter, University of California, Irvine: Were patents exclusive to companies outside the consortia? Was there a university patent agreement?

Matthew Tirrell: Members of the consortia have a right of first refusal for a most-favored, nonexclusive license of patents on inventions made solely by University of Minnesota faculty, staff, and students. If no member company elects to obtain a license, then the patent reverts to the University of Minnesota's Patent and Trademark Marketing operation to handle. This is a quid pro quo for the University of Minnesota's support for the center. All member participation agreements include the university's standard patent agreement.

Andrew Kaldor, Exxon: Was the measure of success for the Industrial Fellows the repetitive business?

Matthew Tirrell: Yes, the principal measure of success was the extent to which members continued to make cash contributions and remain involved, in spite of their own cash-flow problems, reorganizations, and difficulties in designating Industrial Fellows.

David Schetter: Was it really an award without university overhead? Was it really a gift?

Matthew Tirrell: Yes, in accordance with University of Minnesota's policies, these industry contributions were considered as unrestricted gifts.

Richard Alkire, University of Illinois at Urbana-Champaign: Is there just the one category of membership?

Matthew Tirrell: There were two major categories of membership: the sponsor, contributing $80,000 per year, and the affiliate, contributing $30,000 per year. Sponsors can participate in any and all research program areas, whereas affiliates can participate in only one.

James Desveaux, University of California, Los Angeles: You have outlined an impressive degree of innovation and adaptation. A question by another member of the audience has spawned curiosity for me about some organizational design issues, related to the different types of industrial partnerships permitted under the arrangements you described. What you have described sounds a lot like the kind of system that the Microelectronics and Computer Technology Corporation in Austin put in place a number of years ago. In fact, what you have discussed even bears some resemblance to partnership innovations recently made by the partners of the Semiconductor Research Corporation. Where did the ideas come from for the collaborative innovations that you outlined? Did you actively scan other models? To what extent are these models replicable?

Matthew Tirrell: The leadership for developing the proposal to NSF was shared by D. Fennell Evans (who became the center's director following the award) and Robert Stokes (who retired from Honeywell after serving over 25 years as a research scientist and who also served two years as an NSF program officer). Evans and Stokes visited a dozen or more companies and solicited their advice and suggestions. These were incorporated into the ideas developed by key faculty members, who later became research program leaders. Although we were well aware of the NSF's ERC guidelines, we "did it our way" for the most part.

Richard Alkire: What are some examples of how you changed what you are doing? You talked about it in positive terms, but I am curious about specifics.

Matthew Tirrell: This was an evolutionary process in which people and ways of doing things changed in response to opportunities that developed as we interacted with companies, especially through their Industrial Fellows. We soon learned that we could base high-quality Ph.D. theses on industrial problems based on a continuous dialogue between our faculty and students and their industrial counterparts.

Laren Tolbert, Georgia Institute of Technology: I've probably been an administrator too long, which involves indirect costs. You made the comment about faculty supporting all the administrators. But my administration would agree that these are real costs, which they are. William Wakeham talked about a plan in which the full costs were paid, although you never said exactly what your indirect cost rate is.

Are we selling our research too cheap when all we ask for is the dollar that goes to the student and not the dollar that pays the lights or pays the person who writes the check to the student? Therefore, are we subsidizing the industrial research through resources that we have?

Matthew Tirrell: Unfortunately, the typical university perception is that academia has a lot to offer companies, while companies have nothing to offer in return. We established the CIE as a partnership that brought tremendous advantages to the university. These advantages will pay dividends for a long time to come.

Bruce Harrer, Pacific Northwest National Laboratory: I had a follow-on question to that with regard to the indirect costs. There are a lot of centers listed in Figure 4.1. Did they all get a waiver on their indirect costs? Also, for the source of funds in Figure 4.1, is it only the industrial contribution that received a waiver on the indirects, or did all these principal investigator grants get a waiver on the indirect costs as well? Eventually, someone has to pay for the overhead for these centers, and if you grant waivers, you will have to pay for it somewhere else.

Matthew Tirrell: Indirect costs were paid on the NSF ERC grant and on all grants to principal investigators. The ERC and other grants dictate a large number of nonresearch activities (e.g., time-and-effort cards) that companies have no interest in paying for.

Christopher Hill, George Mason University: As you've gone from the original center to the new center, have you noticed a difference in how industry is interested or willing to relate to you? It's getting tougher to get this kind of open-ended, nonspecific support. Is that your experience?

Matthew Tirrell: Upon the receipt of the award for the new Material Research Science and Engineering Center (MRSEC), we structured, in concert with our member companies, the "CIE/MRSEC Industrial Partnership" which allows companies to participate in the activities of either or both centers by executing a standard participation agreement. In their other relationships, the two centers operate independently.

David Schetter: Have you approached your current members for sustaining or increasing any contribution to give you a core; and if you have, what was their response?

Matthew Tirrell: Over the past ten years, there have been two increases in annual fees, agreed to by member companies. As we continue with the new industrial partnership, the same basic fee structure will apply.

Thomas Manuel, Council for Chemical Research, Inc.: To link Christopher Hill's point with Laren Tolbert's for this discussion, we are dealing with a classic perception in academia that things are priced based on the cost for them. But this is not the case. Things are priced based on their value in the marketplace. And companies, due to the dynamics of the globe today, are forced to be very hard-nosed about their own profitability, getting the results for their shareholders at the lowest possible cost.

So the issue is not that you are selling your research too cheaply. The issue is whether a company will pay whatever you charge compared with what, for example, Imperial College, Taiwan, or what a cooperative research and development agreement involving Los Alamos will charge. This is a common disconnect between the academic and the industrial communities on this issue, and we need to be aware of it.

Matthew Tirrell: At CIE, we have been careful in setting the prices we charge for what we offer and have always done this in consultation with our industrial advisory board. Also, our member companies report great benefit that has accrued by their taking the results of our nonproprietary research and applying them in the context of their own products and processes. We have solicited "nuggets" from companies regarding the benefits from their participation in the center. In one case, the company reported savings of over $25 million thanks to the application of a technique learned by one of its Industrial Fellows during a campus residency. In another case, a new product resulted from the application of fundamentals garnered at the center. In all, we have recorded over 365 of these "nuggets."

Thomas Manuel: The problem is that the economics that I described are localized free market parochial microeconomics. In a macroeconomics sense, and in the enterprise, the nation has a different set of values. The timeframes are so long that individual companies find it hard to deal with them.

Matthew Tirrell: Indeed, this is an issue that we often confront. Some companies have simply chosen not to participate based on timeframe concerns. Some companies have chosen to withdraw based on the shortening of their time horizons. Simply stated, companies and their timeframes change quite often, with some companies coming and going.

5

Building Industry–University Research Partnerships: Corporate Perspective

John C. Tao
Air Products and Chemicals, Inc.

Air Products and Chemicals is a $5 billion company with almost 17,000 employees worldwide, operating in more than 30 countries. Our company supplies industrial gases and specialty chemicals directly to other industries, as opposed to consumers, and undertakes about $130 million of R&D, between 4 and 7 percent of which is done externally.

Why do external technology? Box 5.1 gives some of the broad reasons, focusing on the university partnership that Air Products has.

Certainly external technology development is something that wasn't done a whole lot 20 years ago. On the demand side, those of us in industry have to face faster development cycles. We're always driving to get the cost down—cheaper, faster, better. Moreover, everything is becoming more complex. Businesses and relationships are becoming more complex, and the problems are becoming more complex. We can hardly have all that expertise internally. At the same time, industry has reengineered, downsized, or right-sized, which makes it that much more difficult to do more with less.

BOX 5.1
Changing Environment for Research and Technology Development—
Why Do External Technology?

Demand
- Faster development cycle time
- Cost and margin pressures
- Increasing complexity
- Flat and shrinking technical expertise from reengineering

Supply
- Global availability
- Faster and higher growth of research knowledge
- Small start-ups and spin-offs from universities
- Increasing talent pool
- Defense funding shifting to industrial uses

Transfer of External Technologies

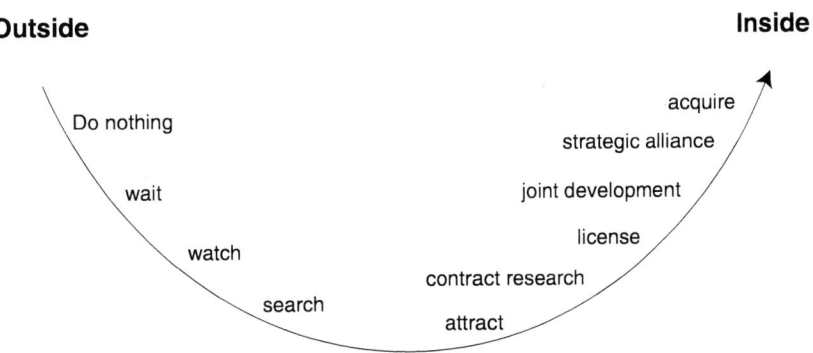

FIGURE 5.1　Technology transfer.

In looking at technology external to the company that you are interested in, there are a number of things you can do (see Figure 5.1). You can start with doing nothing. You can just read about it. You can search around. You can attract some students to have them work or actually have them do the research at the university under contract. You can even acquire the technology or the company that has the technology.

Thomas Manuel commented earlier on the global supply of technology. Universities in the United States are just one place outside of corporations where good science and engineering are done. There is always this increase and faster growth of research and knowledge. There are more and more spin-off companies coming out of the universities, and this talent pool is growing. Furthermore, there has been a shift in the federal government's spending from defense to more industrial-related R&D.

At Air Products we do work at national laboratories. We also work with other companies, both large companies, in the case of Chemical Industry Environmental Technology Projects—a limited liability company that we formed with DuPont, Akzo Nobel—and with a lot of Small Business Innovation Research-type companies that bring a lot to the table from a feasibility standpoint in terms of new ideas. But in the remainder of this paper, I focus on the industry–university relationships and partnerships.

INDUSTRY–UNIVERSITY RELATIONSHIPS

As shown in Box 5.2, the size and purpose of the research enterprise have grown substantially over the past several decades. Research during the pre-World War II era was mostly a search for new knowledge, with a small amount of government funding and very few research partnerships among the different sectors. During the post-Cold War era, however, federal agencies funded most of the research focused on defense and health. The total amount of research conducted in the United States increased substantially, and significant discoveries were made in electronics and biomedical research. At the same time, industrial research grew substantially as well. A few of the "high-tech" industries were born. In addition, the interactions between academia and industry grew, especially at the basic research level.

Figure 5.2 shows the sources of funding for academic basic research from 1993 to 1997. It makes

BOX 5.2
The Changing Research Enterprise

Research in the pre-World War II Era
- Search for new knowledge
- No substantial government funding
- Few industrial science research labs
- Little or no interaction between academic research and industry
- No partnerships or alliances

Research in the post-Cold War Era
- Search for new knowledge and defense, economic, and health needs
- Government is primary funder
- Scientific research is large enterprise
- Explosion of progress (especially electronics, medicine)
- Growth in industrial research
- Growth of technology-based industries
- Increasing interactions between academia and industry and true partnerships

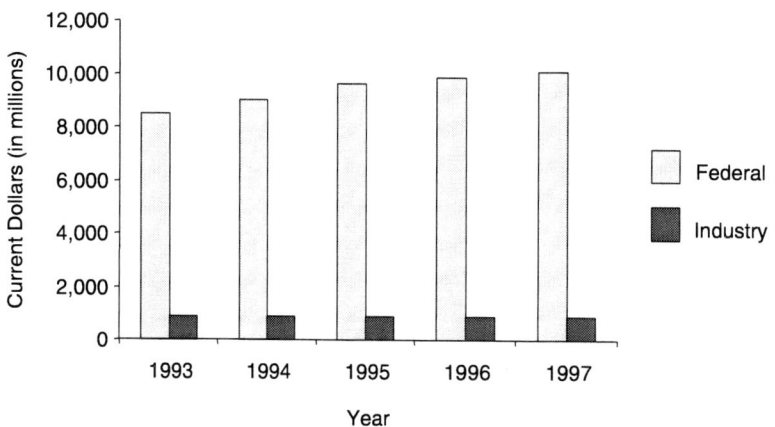

FIGURE 5.2   Academic basic research funding sources, 1993 to 1997.

the point that the federal government is still the major source of the university research money, and industry provides, at best, 10 percent.

Figure 5.3 gives the breakdown of industry funding for academic research ranging from basic to applied to development. It shows that industry does not go to universities to conduct work downstream of research (i.e., development work is not the university's strength). When we have a pilot scale or when we do further applied work, we do that primarily in-house.

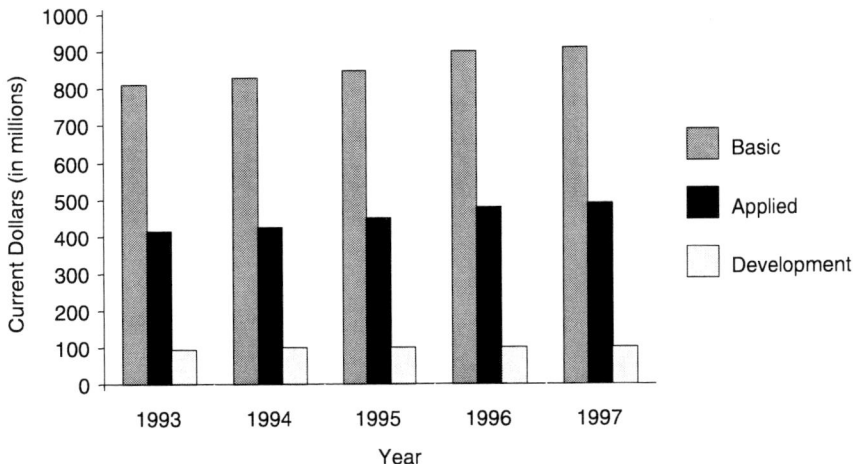

FIGURE 5.3   Industrial funding of academic research, 1993 to 1997.

SURVEY FROM THE COUNCIL OF CHEMICAL RESEARCH

Let me shift gears and talk about the survey from the Council for Chemical Research (CCR) done in July 1997. This was a survey sent to 450 members of the CCR, mostly chairs of chemistry and chemical engineering in the United States, but we also included people overseas in Europe, Japan, and Taiwan. We received almost 200 responses (shown in Figure 5.4): about half of the responses were from the United States and the other half from overseas.

The open squares in Figure 5.4 represent the current situation, and the open circles show what the respondents said there should be in the future. The higher the number, the higher the emphasis, as can be seen when collaboration moves from a scale of about 4.5 all the way to almost 9. The respondents would like to see participation from industry—which is very low right now, close to 2.5—go above 8.

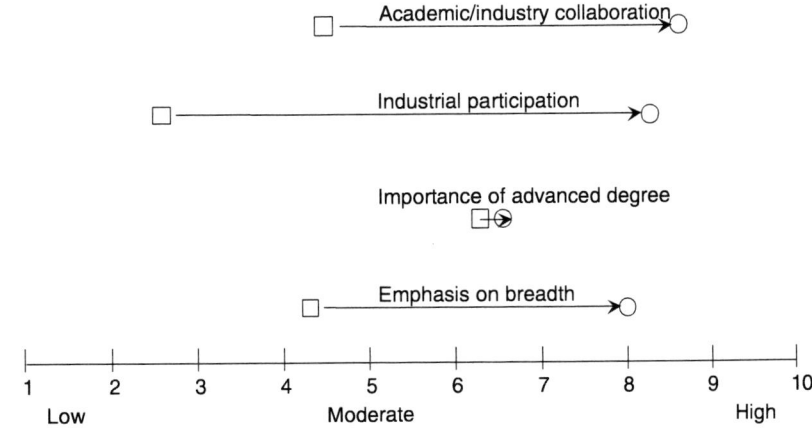

FIGURE 5.4   CCR survey, industrial participation. Respondents judged the current state of industrial participation in education programs to have the lowest emphasis and the greatest desire for improvement.

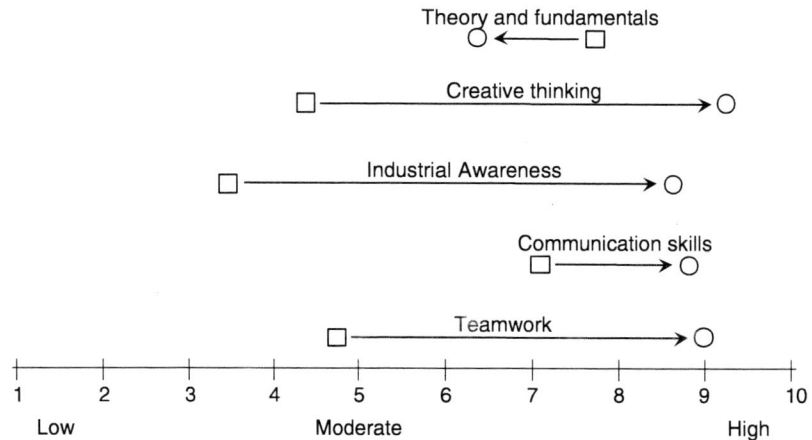

FIGURE 5.5   Graduate education. Sustantial changes are desired in graduate education, with most countries' respondents believing that theory and fundamentals are overemphasized.

The survey also looked at graduate education and how it prepared students for the work place (Figure 5.5). It looked at training in theory versus more practical and applied, and the arrow goes from right to left only for theory and fundamentals. Everything else shows significant moves to a high emphasis on teamwork, a little less than 5 all the way up to 9; industrial awareness is a little over 3 to close to 9.

WHAT EACH PARTNER BRINGS TO THE TABLE

As shown in Box 5.3, the universities obviously provide a lot of research and good work to complement industrial research, and the costs are lower. Industry seeks access to professors as consultants. We look at certain technology rights. We look for new ideas, for opportunities. Of course, we recruit the graduates. And we can look at the leveraging of our funding with the funding from the federal sector.

BOX 5.3
What Universities Offer to Industry

- Fundamental research and enhancement of own research
- Additional high-quality scientific research
- Costs—lower overheads
- Access to consultants
- Patent rights
- New ideas and business opportunities
- Recruitment of graduates
- Leveraging of federal funding

BOX 5.4
What Industry Offers Universities

- Access to world-class and important problems
- Access to other activities of the innovation cycle (development, engineering, manufacturing, etc.)
- Access to a different culture of thinking
- Market awareness
- Enriching teaching programs
- Job opportunities
- Research funding at the margin

Industry, on the other hand, brings relevance (see Box 5.4). It brings some real live problems. University researchers can gain some access to the activities beyond the research they do. They can see how a product finally gets out the door downstream of the innovation cycle. They get access to an entirely different type of culture and thinking. They learn about the marketplace. All of these perspectives should help the universities in their teaching program, and, of course, the students gain industrial job opportunities. Also, the funding is approximately 10 percent when compared with federal funding.

Technology Transfer

There are a number of ways to transfer the technology from a university to industry (see Box 5.5). Much of the research is published. You can hire someone to tell you exactly what they know. You can hire students. You can give them contract research work. You can build an alliance, license the work, buy the spin-off company, or do some work with the spin-off company. There are different degrees of usefulness and different degrees of timing and value in terms of how much you gain.

BOX 5.5
Technology Transfer from Universities

- Publications
- Consulting
- Hiring graduates
- Contract research
- Research alliances
- Research centers and consortia
- Licensing
- Spin-off companies

BOX 5.6
Industry–University Interaction

	IP Value	Timeliness
Gifts	Low	Low
Consultancy	High	Medium-High
Consortia	Low-Medium	Low
Contracts	High	High
Hiring graduates	Medium-High	Medium

As can be seen in Box 5.6 the intellectual property (IP) from a gift to a university is fairly low, and there is no control on the speed of results, whereas with a research contract or work in a consortium, you achieve different levels of timing and a different value.

Partnership Models

I now turn to some examples of partnerships, i.e., the different types of relationships possible in industry–university research (see Box 5.7). Professor Wakeham talked about the Air Products relationship with Imperial College. CCR has a partnership with Pennsylvania State University, and we are about to launch one with Georgia Institute of Technology.

The Center of Interfacial Engineering at the University of Minnesota is an example of the multicompany–single university collaboration model. Perhaps the Massachusetts Institute of Technology (MIT) media

BOX 5.7
Industry–University Research Partnership Models

I. Single Company–Multiple Universities
 — Hoeschst Celanese with Rutgers University, North Carolina State University, University of North Carolina
 — Air Products and Chemicals, Inc. with Imperial College, Pennsylvania State University, Georgia Institute of Technology
II. Multiple Company–Single University
 — MIT Media Laboratory with Hewlett Packard, Phillips, Digital Equipment Corporation, and others
 — Imperial with Air Products and Chemicals, Inc., British Petroleum, Rhone Polenc, Unilever, Fujitsu, Microelectronic and Computer Technology Corporation
III. Multiple Companies and Multiple Universities
 — Pennsylvania Infrastructure Technology Alliance with Lehigh University and Carnegie Mellon University

lab is a better known example. In the chemical industry, the Separation and Research Program at the University of Texas at Austin is well established. There are four multicompany–multiuniversity cases as well. The one that we are closest with is the Pennsylvania Infrastructure Technology Alliance. This involves Commonwealth of Pennsylvania money, and it goes to two former National Science Foundation (NSF) centers, one at Lehigh University and one at Carnegie Mellon University, and there are over two dozen industrial participants.

Issues and Problems

With any two partners, there are always problems and issues to deal with. There are differences in goals and values (see Box 5.8). Those of us in industry deal with the marketplace; we focus on return on investment. We are very cost conscious. We go out of business if we are not cost competitive. We are for-profit; we have to provide a return to our shareholders, and we always have to keep the results of innovation as a competitive advantage. That is how we make money. And, of course, timing is everything.

On the university side, first on the list of goals is advancement of knowledge. Professors like to set their own objectives. They publish. They need to educate students. And the recognition of their work in the academic institution is very important. Their fame equates to power, at least at the researcher level. Timing is not that important.

When it comes to innovation, again there are some differences (see Box 5.9). At the university,

---

BOX 5.8
Differences in Goals and Values

Industry
- Market driven
- Emphasis on return on investment
- Cost conscious
- Profit oriented
- Keeps results and competitive advantage
- Timing is everything

Universities
- Advancement of knowledge
- Academic freedom
- Publication of results
- Education of students
- Recognition of research and institution
- Relaxed time frame and milestones

BOX 5.9
Issues in Partnerships: Innovation

Universities
- Driven by curiosity
- Conceptual breakthroughs
- Discontinuous technology
- Displace technology
- New markets

Industry
- Results driven
- Maintenance of core technologies
- Continuous and incremental improvements
- Capture markets
- Market share of existing markets

innovation is driven by curiosity; researchers tend to make breakthroughs and their innovations often lead to discontinuous technologies or new technologies to replace existing ones. Innovation in industry is more results driven, with the focus on core technologies and incremental improvements. The improvements offer new markets and increase the market share of existing markets.

Publication in an alliance or a partnership is usually not a problem (see Figure 5.6). In most cases we agree on a time period to review the manuscript. We want to have enough time to file a patent and obtain approval or at least some rights to delete any sensitive information. And in a lot of cases there is confidential information involved from the company, and there are typical clauses in any confidentiality statement.

The Association of University Technology Managers (AUTM) does an annual survey of its members. Results of the latest survey as shown in Box 5.10 indicate that total university-sponsored research

- **Period to publish / present**

- **Rights to review and approve / delete**

- **Company's proprietary information**

- **Who is responsible for confidential information**

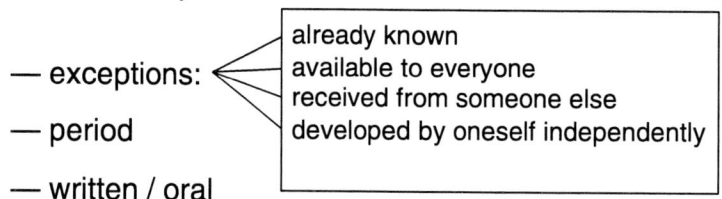

— exceptions:
  - already known
  - available to everyone
  - received from someone else
  - developed by oneself independently

— period

— written / oral

FIGURE 5.6   Common publications-related issues.

BOX 5.10
AUTM Survey (1997)

- Total university-sponsored research—$19.8 billion (+6%)
    $2.2 billion (+15.8%) from industry
- Patents
    — Disclosure: 11,303 (+10%)
    — Filed: 4,267 (+40%)
    — Issued: 2,645 (+24%)
- License income
    — $483 million from U.S. universities
    — 79% from life sciences
    — 74% from top 15 of 132 universities

BOX 5.11
Common Intellectual Property Problems

- Patent ownership and assignment
- Royalty-free nonexclusive
- Option period for royalty-bearing exclusive
- First right of refusal
- Up-front fees and annual minimums
- Patenting costs and maintenance fees
- Right to use all other know-how
- Third-party funding and their rights (especially government)
- Use of confidential and proprietary information

---

is approximately $20 billion, up 6 percent. The percentage from industry is up even more (16 percent). Patents are increasing also, both in filing (+40 percent) and actual numbers issued (+24 percent). For licensing income, approximately 80 percent is from the life sciences, which is really biomedical and agrichemical. Close to three-quarters of the licensing profits are from the top 15 universities out of the 132 that responded.

In fact, a new report done for NSF by Professor Yong Lee of Iowa State University, called "University–Industry Collaboration on Technology Innovation," includes a survey of approximately 700 people, of whom over 400 are faculty members and about 150 are from industry.[1] A question about what faculty members look for in a collaboration was, "What are the personal reasons for getting involved with industrial-sponsored projects?" The answer "getting funds for the graduate students and lab equipment" was close to 70 percent. "Gaining insights into their own research" also ranged close to 70 percent. "Testing the practical application of their own theory in research" scored 65 percent, and 91 percent ranked "to look for funds from industry" as the least important reason for collaboration.

INTELLECTUAL PROPERTY

Some common IP problems are listed in Box 5.11. First on the list of IP issues is "Who gets to own what?" Does the sponsoring party have to pay for it after the patents are issued? Is there an option period to get a license? Does the sponsor get the right of first refusal to an exclusive license? Do they have to pay any up-front fees? Who pays for the patenting costs and the maintenance fees? That may sound like a trivial matter, but it is not when you consider multiple applications and foreign patents. For some foreign patents, just the maintenance fee alone could be $15,000 for a few years. In regard to the rights to use the other know-how that came out of this work, what are the rights of the third party? And who gets to use what information afterwards?

To answer some of these questions, we need to look at the possibilities of both sides. What are the possibilities for the universities if the company owns the IP? (see Box 5.12). There are three possibilities

---

[1] Young S. Lee, "University-Industry Collaboration on Technological Innovation: Views from the Ivory Tower," *Policy Studies Journal*, 26(1): 69–84, 1998.

BOX 5.12
Company Owns Intellectual Property: Possibilities for University

Rights
- No rights without permission of company
- Equal rights and free use for own purpose; negotiated rights for third parties
- University can use for research; no rights for third parties

Benefits
- No reward for commercial exploitation
- Subsequently negotiated reward
- Preagreed upon reward

---

on the "rights" and three possibilities on the "benefits." The university cannot get any rights without the permission of the company, can have equal rights, or can get some kind of limited rights to use it for research, but cannot do any licensing. And there are a range of benefits. There is no reward, some reward, or something that can be negotiated afterward or beforehand.

When the university owns the IP, the company then has to pay for an exclusive license, pay for an nonexclusive license, or get a royalty-free, nonexclusive license (see Box 5.13). And again, you go through the same benefits on the sharing of royalty and licensing income from the third parties.

Perhaps Figure 5.7 best summarizes how I look at the situation. It is really not black and white. I gave the two extreme cases above. One party owns the IP, and you have to cut a deal with the other party. The model to look at, from the type of partnership to the type of contractual relationship, asks the questions: How much is the funding, is it 100 percent, is it partial, how much of it is basic, how much of it is applied, and how much input and background knowledge does the company bring? Is there a model in which they just sign the check and do not get too involved until the results are out?

In Europe, some of the universities will charge a fee. This gets to the issue of overhead. They actually make money. A decision in that case was, "Okay. We aren't going to argue with you on intellectual property. Who knows whether anything will come out of this research? Who knows whether there's going to be an invention? You pay us a fee, right up front, versus uncertainty later, and you get to own the IP. We know exactly what that's worth." Very few U.S. universities ever think about that.

BOX 5.13
University Owns Intellectual Property: Possibilities for Company

Rights
- Company pays for exclusive
- Company pays for nonexclusive
- Company receives free nonexclusive

Benefits
- Company receives preagreed upon reward from third-party licenses
- Company receives subsequently negotiated reward
- Company receives nothing from third parties

61

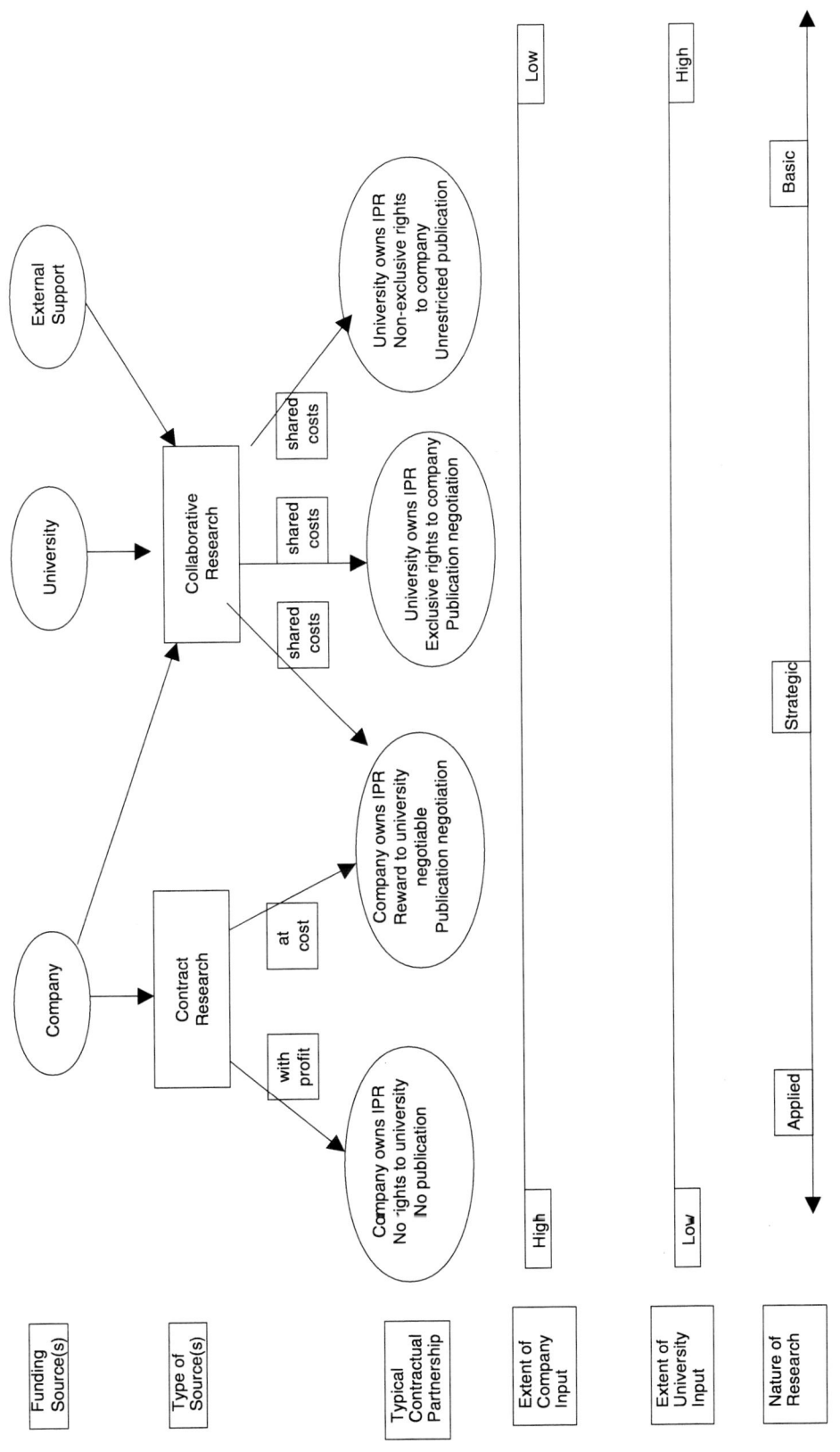

Funding Source(s)			
Type of Source(s)			
Typical Contractual Partnership			
Extent of Company Input	High		Low
Extent of University Input	Low		High
Nature of Research	Applied	Strategic	Basic

FIGURE 5.7   Spectrum of arrangements for university–industry partnerships. IPR, intellectual property rights.

BOX 5.14
What Makes a Good University–Industry Relationship?

- Good historical relationship
- Complementing and overlapping strengths in core technologies
- Clear goals and roles
- Good teamwork
- Good communication
- Good science
- Mutually benefiting (share success and failure)
- Intellectual property ownership agreed to

BOX 5.15
The Bottom Line

- Dollars in research contracts
- Facilities, equipment, and materials used or provided
- Background technologies used
- Expertise provided by both
- Time committed by team
- Timely decisions
- Recognition of people and organization
- Meet each other's expectations

The contract research with full overhead is another case. When you have cost sharing from a third party or, in most cases, with a federal agency, then Bayh–Dole kicks in. The university owns the IP rights, and you have to negotiate whether you get a nonexclusive, royalty-free license or if you have to pay for that license whether it is exclusive or not.

Another trend that I believe will be increasing is to reverse the IP transfer. Increasingly companies will be giving away IP to the university for the university to exploit. And the company will get a tax benefit. We are going to see more and more of this.

What Makes a Good Partnership?

First, it is important to have a good historical relationship (see Box 5.14). Professor Wakeham talked about the Imperial College partnership with Air Products in which a former CCR vice president of engineering, an Imperial College alumnus, brought us together. CCR now has over 500 Pennsylvania State University graduates on campus. These alumni help.

It also helps to have overlapping strength in core technologies, the kind of work industry does and the kind of research that professors are interested in. It is important to have very clear goals, good teamwork, and good communication. You must always do good science. At the end of the day, that is the only result you get. You have to share the benefits, and you have to work out the IP issues ahead of time.

What I learned is that the more you put in, the more you get out (Box 5.15). If all you do is write a check and wait for the results to come a year later, that is not going to work well. You can measure the amount of dollars you put in at an institute; you can look at the amount of facilities, materials, and the equipment you have built; and you can look at how much background work you put in, how much expertise the two parties put in, the commitment from the teams, and whether decisions are made quickly. Do you recognize the people when successes are realized, and, most importantly, do you meet the expectations of the parties going into the partnership?

DISCUSSION

Henry Kohlbrand, Dow Chemical Company: Would you comment on the direction that your university relationships are going? Do you see the number of institutions that you are actively working with decreasing in favor of a few strategic relationships, or do you see a blend of strategic relationships and one-of-a-kind agreements?

John Tao: It's the latter. We never ask our researchers to only work at Imperial College or Penn State or in the future at the Georgia Institute of Technology. Traditionally we have worked with about 40 different institutions, including those overseas. And we see those partnerships continue, but we built a few alliances that we are also committed to. And as Professor Wakeham said, these are long-term relationships, but they are not everything.

David Schetter, University of California, Irvine: You mentioned that 4 to 7 percent of your budget was for extramural activity. Do you see that going up in the future as was projected earlier?

John Tao: Yes, for the reasons I stated: the drivers on the demand side as well as the supply side. We cannot possibly do everything in-house.

Ashok Dhingra, Dupont: I found some things in your presentation that are common to what we are doing at Dupont. But in your case I can see that it was a prior relationship between your vice president of engineering who had some connections, which helped start the project. Do you give a certain amount of money to a university and then build projects based on that funding? In the case of Dupont and many other companies, the corporate funding is not there anymore as it was in the past. How do you get the businesses to participate in funding these programs?

John Tao: A typical project or proposal coming in from a university has a staged approach. In the first stage, it may be very exploratory in which the feasibility has not even been established. Then it's fair to ask for corporate money to help get the project kick-started. Very quickly, once we move beyond feasibility, let's say, to the second year, we'll ask the business units to kick in at least half the money, and hopefully beyond that they take over.

Ashok Dhingra: Do you see continued support in the areas of restructuring and reengineering and the changes in management? In other words, you start a program and then you want business to participate in it. Do you see this happening?

John Tao: Yes, and we have not had any problems.

Fritz Kokesh, Massachusetts Institute of Technology: You mentioned that you have had some experience in which European universities will allow ownership if the price is greater than their cost. Could you expand on that? Is that across-the-board markup or is that on a per patent basis? How does that work?

John Tao: In that one case at a university in Holland, it was 20 percent extra overhead, but it's not per invention.

Christopher Hill, George Mason University: Earlier I mentioned that the tax status of the university could be a problem for that scheme in the United States. You said that you had been able to do it here. How do you get the work done?

Janet Osteryoung, National Science Foundation: What is this tax problem?

Christopher Hill: Universities are nonprofit institutions. However, income from sports events, book-stores, and cafeterias may be taxed as unrelated business income. It has been said that if universities start to act as a contract research shop, particularly if they seek to earn a profit, it may put the tax-exempt status at some risk across the board. At least universities have been cautioned about this.

David Schetter: There is the Tax Reform Act as well, which means that universities are actually legally precluded from prevaluing technology that has not yet been invented if the research was conducted in facilities built with tax-exempt bonds. It's the law. So you would love to do that sometimes, but it depends on whether the research is actually conducted, and it will challenge the tax-exempt status of a university.

John Tao: I am not sure that I'm the best person to answer that question, but I know some people have done it with a separate subsidiary company.

Christopher Hill: Many of us have created various subsidiaries to get around this problem. If you create a profit-making subsidiary, then you can do those things.

John Tao: The American Chemical Society (ACS) has a very profitable subsidiary, Chem Abstracts, and they are a 501(c) 3 organization.

Frank Feher, University of California, Irvine: When you do research at a university, the research is usually done by graduate students or postdoctorates who have a finite time at the universities. So you start to develop what could be a pipeline problem if you are going to be doing your research with essentially what are temporary workers. To maintain a flow of workers into the system, you must have prospects, good economic prospects, for workers at the other end. Would you care to comment on your model for research in which 80 percent of the research conducted at a university has some connection to industry? Are there going to be positions available for the people doing the work as temporary researchers, and can we keep the process going?

John Tao: Well, the professor is still there.

Thomas Manuel, Council for Chemical Research, Inc.: I heard the question differently. The issue is if research is funded at universities at an increasing rate and if the workers are people who are getting their degrees so that they can come out of the university and do research somewhere else, then the research that industry was going to do in their own labs is now done in the university. If so, do you have a material balance? That's the old question of supply and demand of technical people. I have two observations: First, the postdoctorate pool is the surge tank on this in the short term. And second, if I recall, when Ned Heindel was ACS president, ACS did a big study of all this and found that everyone got a job somehow. What happens is that people who are in industry turn into salesmen. If they are not very successful, they become managers. And then finally they retire and work for nonprofits. So there are other avenues out of this great tank of people. In theory, it's a problem, and someday it may happen, but you have to ratchet up the ratios before this becomes very serious.

Andrew Kaldor, Exxon: Do you have any examples of the type of reductions that you are able to achieve through university interactions?

John Tao: I can't give you any specifics.

Joseph Gordon, IBM: You implied that you had a sort of sliding scale for ownership of intellectual property and the amount that you are willing to pay for it. Do you actually have a range? And what fraction of your extramural research is in the form of a gift? What fraction is through contracts in which you own everything?

John Tao: Let me put it this way. We have very few in gift form, and we have more on the contract side of it.

Todd La Porte, University of California, Berkeley: You have gone through quite a long experience yourself on this. Could you talk about the surprises you've had in this experience?

John Tao: Some of the problems are people issues. It has nothing to do with good science or good research. We had to deal with one case in which the communications broke down, and we finally realized what happened: There was going to be a publication, and it was a battle about whose name came first. But very rarely do we have difficult problems that cannot be solved once you get beyond the people problems.

Todd La Porte: I didn't mean just negative surprises.

John Tao: Yes, we have had good surprises. We had a project in which we were looking at specific interactions of gases in the electronic etching process. There was an unexpected result in that the addition of some CO_2 gave a much better life to the tools. And there was an invention that no one expected. That is translating into dollars for us right now.

6

Diluting Public Patrimony or Inventive Response to Increasing Knowledge Asymmetries: Watershed for Land Grant Universities? Reflections on the University of California, Berkeley– Novartis Agreement

Todd R. La Porte
University of California, Berkeley

INTRODUCTION

In the context of this workshop, my presentation takes a somewhat different approach. We all share a strong interest in helping universities to cooperate with government and industry. Such cooperation, when carried out under the right conditions, will provide substantial benefits to all concerned. I approach this problem from the other, nontechnical part of the university, and my comments are an attempt to show how the rest of us are affected by research partnerships in the area of science and technology. I am a political scientist, a real outsider here who comes from a long distance substantively to the issues that concern partnerships with industry and government. What follows is a cautionary tale that broaches some unsettling issues. I hope it might begin a path toward developing an institutional matrix within which to engage industries while minimizing the distortion to the values of public institutions and public universities in the United States.

BACKGROUND

In 1996 and 1997, the College of Natural Resources (CNR) at Berkeley, and especially its scientific department, confronted the old problem of an eroding research infrastructure, accelerated in this case by a decline in state legislative support for the University of California (UC) system, down to about 30 percent of our total resources. There has been a relative surge of capital investments on the campus in the past couple of years, which has been mainly for seismic safety upgrades designed to improve safety for people escaping from buildings during an earthquake. But this does not protect the buildings themselves. If we have a significant event, we are likely to be able to get out, but not be able to return. So the upgrades that we had thought were two-way streets, if we have to use them, may cause a very interesting research continuity problem.

The new twist to the shortage of research funding, and the one that prompts this presentation, is very important, and in a sense is new and novel, at least on our campus. This is the growing recognition that in some areas, and particularly the agricultural genome area, four or five mega international corporations—the so-called "life sciences corporations"—appear to hold as proprietary information a key fraction of the field's salient intellectual capital. They maintain this information so closely that our Department of Plant and Microbial Biology (PMB) could no longer assure the university that it could provide a top-flight education for the graduate students. There is a sufficient monopoly over this kind of intellectual capital, and the department encountered a series of surprising outcomes over the past four or five years. This is a very interesting and troubling consequence. One might suggest, to be optimistic about this, that we in the higher education world have been very successful in producing our own competitors in the private sector.

The consequences for the free flow of information are very troubling, and they comprise one of the drivers to the events that have unfolded at Berkeley. The combined concerns of a funding shortage and nonavailability of proprietary data were serious concerns to the people involved. First, there was inadequate laboratory space for continuing the department's outstanding research program and the fear that the department's standing as a major research and teaching resource in the discipline would be significantly diminished. Second and more novel was a fear that the equally important public function of assuring broadly available basic knowledge might be truncated in favor of the self-interests of multinational corporations. It could be argued that the interests of these groups might not be consistent with the economic and agricultural interests of the American West.

The agricultural subplot involves the deep division among students of agricultural resources about the respective roles of aggressive manipulation of genetic materials in the interest of sustainable food supplies on one hand, and other less intrusive methods of sustaining agricultural production on the other. This division could provide a very important disagreement with regard to the colleagues who might be joining this enterprise.

PARTNERING

The challenge at Berkeley was to increase resources both for expensive genetic research and research infrastructure and, at the same time, crack the hold on proprietary genomic information held by the several multinational corporations in the United States and Switzerland.

Gordon Rausser, the dean of the CNR, gave considerable thought to this problem of how to work with the corporate world. He is an economist, and this is partially reflected in the kind of design criteria he used to frame an invitation to the five corporations to bid for the services of the PMB. (See Box 6.1.)

One should realize that "in the public interest" (see premise 3 in Box 6.1) does not mean research on issues that the public thinks are important. It is research—often basic research—for which the outcome is expected both to benefit the public and be accessible to it. Rausser uses the language of "public-good research" as something that comes out of a special subdialect of economics. So the CNR had this interesting idea of attempting to deal with a number of issues discussed at this workshop.

In the spring and summer of 1998, the CNR worked through these very interesting principles quite rigorously in the way the college presented them to the potential bidders, and then waited for returns. The four U.S. corporations responded to the competitive bidding with bids that were considered predatory and not at all in the spirit of the design; these bids were rejected, almost out of hand. Novartis, then a new corporation, formed by the merger of two other companies, responded in the spirit of the offer, and the resulting contract worked its way through the legal and administrative offices of the UC Berkeley campus during the summer. In early fall, the dean invited the views of the academic senate, which has a very strong role in shared governance. My abrupt acquaintance with this story began then, as chair of the senate's committee on research. This committee's main role is to dispense small enabling

---

BOX 6.1
Design Premises for CNR–Corporate Relationships

1. Select a *single industrial partner* for a strategic alliance that will maximize the financial, technological, and intellectual benefits for the department, the university, and California agriculture.
2. Use traditional business models to *encourage competitive bidding* among the candidates for partnership.
3. Insist that *the strategic alliance generate large unrestricted funds* for the department to use for research in the public interest, without oversight by the industrial partner.
4. Require that *the industrial partner make significant intellectual contributions to the university* in the form of access to technology and data useful for departmental research that would otherwise be prohibitive because of cost or proprietary reasons.

---

grants to faculty, but is the only locus where research policy matters can be considered by the faculty. The senate was surprised, indeed, taken aback as the details of the draft contract were outlined.

The Novartis–Berkeley Contract

Some of the key provisions of the Novartis–Berkeley contract are outlined in Box 6.2. The PMB in the CNR was to receive $25 million over five years for unrestricted, largely faculty-allocated research support. The spending was to be overseen by a steering committee that has some Novartis representation. About 33 percent of these funds would be reserved for overhead, including renovations of space negotiated with the deans and the administration, support for graduate programs, and overhead for the university. Notably, this is an unusual distribution of overhead in the UC Berkeley context.

According to the contract, Novartis would provide access to its agricultural genomic database on a confidential basis. A $3 million Novartis facility nearby would have workstations to distribute the proprietary data and house several Novartis staff members who would provide technical support for using the data. An original proposal for three adjunct faculty positions for Novartis scientists was subsequently withdrawn, although the university makes this arrangement frequently and it was not a particularly controversial matter. Adjunct appointments go through the regular quality control processes to which all faculty appointments are subject. Finally, the contract would provide faculty and graduate students with access to Novartis personnel on an informal basis. There are some 33 members of the department, and 30 of them signed up for this arrangement. So there is a very interesting concentration of attention that has resulted.

What value does Novartis receive from the arrangement? The contract gives Novartis the first right of negotiations to acquire a percentage (now set at about 30 percent) of the intellectual property from discoveries that may result from the research funded under the contract. The right of first negotiations allows the university to say no—at least in principle—if it is not offered fair market value, so it is not as limiting as many of these relationships. Novartis would also be able to license commercially viable discoveries at their own expense, and Novartis scientists and staff would have informal access to faculty and graduate students.

BOX 6.2
Outline of the Current Contract: Respective Expectations

University (CNR) Expectations

1. $25 million—$5 million for each of five years—for unrestricted, largely faculty-allocated research (steering committees with some Novartis representatives).
2. 33 percent of annual funds to overhead distributed to renovation, graduate program support, and general campus overhead support.
3. Access to Novartis' agricultural genomic database (on a confidential basis).
4. $3 million for a nearby Novartis facility with workstations to distribute proprietary agriculture genomic data and several Novartis staff members to provide technical guidance.
5. Access for faculty and graduate students to Novartis personnel on informal basis.

Novartis Expectations

1. Right of first negotiation to acquire a percentage (now approximately 30 percent) of discoveries that may result from research it helps to fund.
2. Authorization to license commercial-viable discoveries (at their expense).
3. Access informally for Novartis scientists to faculty and students.

There is, as well, an array of patent and legal safeguards. Among the most important is a provision in the contract that termination can be initiated by either party with a year's grace.

Governance and Oversight

The faculty had also become interested in the governing structure, which is detailed in Box 6.3.

A six-member advisory committee oversees relationships between Novartis and the university generally. The university representatives are the vice chancellor for research, who engineered much of this; the dean of the CNR, who was the primary engine behind this; and a member of the faculty, unaffiliated with anyone involved with this work. The unaffiliated role was added at the insistence of the faculty senate. It was originally to have been a faculty member from the CNR, but the agreement was amended because the senate believed that such a faculty member would not be sufficiently uninvolved.

Novartis has three people on the advisory committee, too: the president and CEO of the Novartis Agricultural Discovery Institute (NADI) and two co-presidents of the Novartis Agribusiness or Biotech Research Corporation. There are also two nonvoting ex officio members, the contract principal investigator (PI), who is the chair of the research committee, and the chair of the PMB. This makes sense in terms of providing data and information about how things are going.

Managing the day-to-day affairs is a five-member research committee that allocates funds among the PMB faculty. This committee has the responsibility of distributing the $5 million every year—actually $3.3 million without overhead.

BOX 6.3
Current Contract: Governance Structure

Advisory committee overseeing relationship between Novartis and the university—
six members, plus ex officio
- Three university-related members:
  Vice chancellor for research, dean of CNR, and unaffiliated faculty member
- Three Novartis-related members:
  President and CEO, Novartis Agricultural Discovery Institute (NADI); two co-presidents, Novartis Agribusiness or Biotech Research Corporation
- Chair of research committee (contract principal investigator) and chair of Department of Plant and Microbial Biology (PMB) as nonvoting, ex officio members

Research committee managing day-to-day matters and allocating funds among PMB faculty—
five members.
- Three PMB faculty:
  Contract principal investigator, as chair, and two PMB faculty selected by department faculty
- Two Novartis representatives:
  President and CEO, NADI, one co-president, Novartis Agribusiness or Biotech Research Corporation

---

The Faculty Response

Once the contract was settled, but before it was signed—scheduled for that fall—the dean talked candidly with senate representatives. Considerable distress and concern arose, within the CNR itself and among members of the senate. There was considerable concern on the campus—a remarkable situation that almost never happens in connection with research relationships.

How did the PMB respond to this? They took umbrage to the reaction of the faculty senate. They wanted to know what was new about this style of collaboration. It involves a PI. It has sponsored research. It is similar to an arrangement. So what is the big deal? Why should the faculty senate have anything to say about the free conduct of research? This sentiment was shared by many of us on the faculty, and the senate had a sense of their insult. Indeed, such an official or informal faculty question is rare to nonexistent on the Berkeley campus.

The primary issue under debate—an issue of great importance at UC Berkeley—is maximum freedom for individual investigators. Everything on campus—however irrational it may seem—is rational when observed under that criterion. And from the department's view, freedom for investigators was essentially what the contract delivered. The contract provided more freedom in allocation of research support than almost any type of contract or grant, and the funds were to be disbursed by their own department.

But the sentiment of the department faculty was not shared by its own colleagues. There are five departments in the CNR. When the faculty returned in the fall and learned what was going on, many of the CNR faculty were distressed for a variety of reasons. One was process: A dramatic change seemed to have been implemented very abruptly, and they did not know what the details were. Although a college faculty meeting had been held on this topic and a web page had been posted, the faculty believed that they had not been informed. And there was another anxiety—anxiety about institutional integrity. The faculty committee on academic freedom began to wonder about what was involved here.

Public universities in California—and particularly for land grant institutions—have a constituency that pays strict attention to what the universities do. This results in intense scrutiny and frequently unequivocal judgments. In other words, one can expect such scrutiny to be applied to the proposals for university–industry interactions that we are discussing here. There was a worry about the relationship with industry versus the growers and people of rural California. People realized that this would be a contract with a Swiss firm, and Switzerland is a long way from the Central Valley of California. Novartis has a very strong presence in California, south in San Diego, and north in the Bay Area. The company has been very clever and strategic in how it has invested in California, and I do not mean that in an ungenerous way. But UC also has had a long history of involvement with California agribusiness that has been quite checkered. Almost always, until very recently, it has benefited the large economic interests in the Central Valley, and this has led to a small political movement on sustainable agriculture within the university and the legislature.

Finally, there is the antipathy to aggressive manipulation of genetic materials. That was not why the academic senate was concerned, although we understood that others on the faculty might be. And there were novel properties in the proposed arrangement with Novartis, including its magnitude and its long institutional timeframe. It also differs clearly from the more common arrangement involving a consortium of 6 or 12 companies, each providing some money and dealing with a small number of researchers (much like the situation at the Center for Interfacial Engineering at the University of Minnesota—see Chapter 4).

UC Berkeley and the UC system have experience at the other end of the continuum, with large national laboratories such as Sandia, Lawrence Berkeley, and Los Alamos. One of the things we learned to deal with is how to manage secure and proprietary information—to be sure it is in the interest of national security. But we do not know how to deal with an intermediate situation of the presence of a large multinational corporation in our midst that holds proprietary data with interests beyond those of the university.

Interestingly, the academic senate's response was much more reserved than that of the faculty in the CNR. The more the senate leadership thought about this—and we had long conversations with a number of the senior committees and the academic senate council—it deepened our concern about what the university was getting into.

This is the origin of the title of this chapter—because we began to ask, How far are we down the path of privatization, of essentially diluting the public patrimony of an intense institutional investment for the public in the interest of some part of that public? Is this tension that we all feel warranted, or is this situation really innovative? Did Gordon Rausser produce a design matrix that addresses only the first example of university–industry cooperation at that proposed scale? What happened at Berkeley certainly will not be the last of these temptations of higher education.

Faculty Concerns

A range of concerns bubbled up spontaneously from the faculty across the university as news spread about the Novartis contract. Much of the concern—expressed by e-mails, phone calls, conversations with colleagues, and other approaches—was directed to the faculty senate Committee on Research. This explosion of faculty concern was an amazing outpouring on the part of a faculty that usually focuses on its individual research and teaching activities.

The Committee on Research forwarded these questions to the vice chancellor for academic affairs, the vice chancellor for research, and the dean of the CNR. We hoped that getting these questions and concerns in the open would allow the administration to explain the benefits of the collaboration and why it is not diluting public patrimony. But the responses from the chancellors' offices were defensive and evasive and were not appropriate for public dissemination.

Questions Raised

An important group of the questions raised by the faculty at UC are general to university–industry collaboration[1] and are included in the attachment to this chapter. As universities heavily embark in partnering with industry, these questions are the kind of issues that will be of concern to those faculty who are not immediately involved in the collaboration. Hopefully university administrations will become apt at answering the questions.

I will highlight several of the questions, many of which are really quite straightforward and contain no surprises. But there is a pattern that gives a sense of the symbolic frame of reference and the enormous ambivalence that is felt on campus.

To what degree was the agreement signed with a full understanding of the implications for the public image of both the college and the university? The university has made several attempts at trying to deal with bioengineering firms. They have all failed rather badly. There is a kind of residuum that I do not know well, but it has not been a positive history. This leads one to wonder: What will happen if we become associated with this kind of research?

The context at the time was interesting. Monsanto and the U.S. Department of Agriculture had recently announced that they were about to license a kind of seed that would not reproduce itself—the so-called "terminator gene." It was a great idea for marketing, but the farmers were a little unsettled because it seemed that one company might control the entire market. People on the campus were saying, "Is that what it means to be involved with California agriculture? You and the people like you are doing the terminator genes?" That did not sound like it was in the public's interest. Unfortunately for the internal dialogue that ensued, this story came out about the same time that we received the UC–Novartis contact proposal. And it sent some of the younger faculty and some of the students of the college into a tight political orbit of remarkable intensity.

I have been asked, "To what degree are public UC Berkeley funds being used either to support or provide infrastructure for research funded by industry?" Others have suggested that the patrimony is being put out for advantage to specific groups on the outside world. But what will be the consequences? What are our requirements? And what is our duty—as a public university—to evenhandedly develop information about the world, the national and social world, for all the citizens of California?

Another area of potential concern is the stipulation of the right to license 30 percent of any discoveries. The contribution from Novartis will increase to the point that this contract will constitute 30 percent of the department's total budget. The department has many other resources, as well, from both the private and the government sectors, so this is a cautionary concern. Novartis is not the whole thing, but it is the biggest thing. Novartis also funds some directed research and is certainly free to increase

---

[1] The attachment includes approximately 50 questions that were received from many corners of the campus. The list has numbered questions that came from the academic senate. The starred questions came from the executive committee of the CNR. The original questions were stated more in terms of the rhetorical "won't this or that happen." They are now framed closer to data-gathering devices.

support for that kind of activity within the same department. The question has become: What will one entity, in terms of its attention-getting capabilities, gain from providing 30 percent of the total budget, and probably half of the rest of the budget in terms of technical intelligence? Consider what the Novartis people will be able to learn and exchange information about when they are on site. It is very strategic in terms of exchange of information about important things.

Returning to the list of concerns, there is a predictable set of questions on institutional development; that is, what is the effect of such a large-scale activity on graduate and undergraduate education. To what degree have there been resources set aside to support unrestricted graduate fellowships—to be allocated by the faculty who are not in the Novartis agreement? The question has come up in a number of different places on campus. If you are going to sell yourself to a strong market sector, how do the rest of us benefit?

The argument takes the following form: When you are considered to be one of the top departments in the country, it is not just the one department that is being evaluated. Everyone around you—your university neighbors—is implicated in that judgment. Departments do not exist in isolation. If the rest of the university were of lower standing, you would not have that top position. It is a judgment on the institutional collective that people are making about your capacity as an analytical resource. So we need to think about not just the immediate self-interest of the department and graduate students, but of others as well who are a part of the institution's world.

When these matters involve making liaisons with private interests, the question arises as to how much of your attention is turned to those private interests before you are no longer squarely operating in the public's interest as in public university. This is a really interesting problem. It has never come up before on the UC Berkeley campus. It is part of the larger frame of reference in that more and more often we are being asked as faculty to partner with people outside our normal environment. One must—or at least should—ask how many of these interests we are being promised to. How many can we join before our capacity to be loyal to our institution and our public mission is seriously compromised? I am astonished at the strength of consensus among my colleagues on these issues, although we rarely spoke about them previously. It turns out that we never realized how much we cared until the Novartis matter arose. It has been an extraordinary catalyst to discovering that we had a community of interest about unspoken subjects.

The next question is a related concern expressed by the college faculty. To what degree has there been a perceived increase in the cultures of "haves" versus "have-nots"? Holders of research grants and contracts are the "haves." I happen to be a "little have," but many of my colleagues are "have-nots." Nevertheless, they are a part of our collective success. The concern here is not just the Novartis episode, but also the other issues that increase the sense of distance among our colleagues. The different approaches and attitudes of various disciplines may create friction at times, but all the disciplines are important.

Another concern is: To what degree has the freedom of association or research, teaching, and publications been restricted formally or informally with regard to this arrangement? This problem is not specific to the Novartis relationship, but it is one in which the faculty has a great interest. When our committee on academic freedom encountered this question, they asked, What are the conditions that would assure these freedoms, not in the first two or three years—that is not a problem—but in the fourth year, the fifth year, and through a second Novartis contract? It is the long-term erosion of these freedoms that concerns us. There is the possibility that the questions one asks will change—with the consequence that we will only ask about what is interesting to industry. While I understand this—it happens to me sometimes too—think about how that sounds to outsiders observing us. What is the level of independence here? Do we have to be concerned? I do not know whether we have to be concerned or not, but one can imagine the possibilities and manage the situation to avoid compromising this essential academic value.

The last set of questions has to do with governance of the institutional evolution, "provisions for exiting/terminating the relationship." In a relationship this exclusive, we do not feel assured by the stipulation allowing formal termination at any time. Although the desire to have this provision is understandable, to think that it provides a formidable protection against institutional erosion is naive. When you have an association with a partner of that scale, with that many people, and for that long a time, it is not reasonable to just terminate the arrangement.

There is a high degree of skepticism among faculty on campus who have been in this situation before. When resources for the university from the state legislature dropped from 60 percent to 30, our research capacity provided the flexibility needed to compensate for the loss. We have a number of faculty who have been in positions of heavy dependence on external funding. They understand what can go wrong and what can go right in relationships with outside funders, and they are the people who are saying that terminating the contract as a significant protection is not credible.

Another issue is prompted by the present governance structure. To what degree is this structure seen as being composed of people who themselves will benefit? It could be imagined that everyone in the department benefits except the few, unaffiliated faculty members. This is a potential source of trouble if you want your governing arrangement to be a good source of early warning. The structure we have is not a very good one. But it is the structure we have and it is the structure we are going to be watching.

One question is associated with an interesting set of issues not previously mentioned at this workshop. "How have plans worked out for developing a research capacity, not only to develop new agricultural products and methods, but to provide insights into the consequences of their widespread deployment?" A relevant example is an experience at the UC Davis campus, where they developed a machine for harvesting tomatoes in great numbers. Traditional tomatoes were too fragile, so the tomatoes were changed. Great fields of these tomatoes can be seen in the Central Valley. These plants are wonderful for ketchup, but not much else. These machines were built in a way that required large plots of ground to make them economically efficient.

Can you imagine what happened to the Central Valley after that machine was invented? The land pattern was transformed from many small farmers, with small land holdings, to a few very large tomato growers using those machines. Who could have known? Perhaps anyone who thought about it could have known. And the university was taken to court about seven or eight years ago for not letting anyone know about the potential consequences of this invention for the society in which we live. The university prevailed in court on a technicality, but think of the residuum of suspicion that people hold about this.

Much of the research discussed in this workshop is intended explicitly to learn how the world works more effectively—to change the conditions that people who live in our society will experience. But suppose we are wrong about the underlying benefits of the technical processes that we provide for others. Are we at the university responsible if something goes wrong out there in the farms? What if we have a runaway terminator gene—however improbable—or something like it? Would we be held accountable? I think we would be if something harmed the wine industry. We must think about the long-term consequences and follow through on our obligations to report those consequences.

My last point concerns the processes that are put in place to involve stakeholders in receiving information about Novartis-based development and seeking advice for research directions themselves. "Stakeholders" in this context are the general public—people without direct financial investment in the program. I contrast this with "special interests" who do have a financial investment. In this situation, a public university can be perceived, if it is not careful, as working for the benefit of special interests rather than the public. I believe the political environment of UC and of higher education in general is more intense in California than in other states. Here we have an interesting dialogue, and it is in that context that we must view the Novartis contract.

Outcome

After all the deliberation, the faculty decided to suspend judgment, recognizing the case as an experiment in progress. This redefined the challenge and allowed us to say "Let's do the experiment right. If anybody in the university or higher education communities should be able to do it right, we're the ones!" What this meant was that the senate would not say yes or no for five years—until the experiment is completed.

This means that the administration, Novartis, the academic senate, and the CNR will conduct a study. Some argued that we could not call it an "experiment" because we had no control group. And it could not be an "evaluation" because evaluations lead to finger pointing and blame. So we are going to call it an "observational following." This "following" is intended to discover and understand the consequences for the institution while attempting to manage the overall arrangement. The questions in the attachment are all hypotheses, framed as questions for which data might be gathered. But we are not just studying ourselves; we are surrogates for every land grant institution that has the same problem. The political and institutional challenges are the same across the country.

The tensions are already present. A survey was done in the CNR in parallel to the senate's effort, and it turns out that two-thirds of the CNR, and many others in the university are focused in hostile fashion against that department. This is not a healthy situation. Somehow, in a collective way, we have overridden the norms of individual freedom of research on our campus. It is astonishing that this kind of sentiment has begun to crystallize on campus. I do not know what the outcome will be, but my committee is attempting to put together a request for proposals to get someone—an external group—to follow the evolution of this institutional development at Berkeley.

Although I have had a number of invitations to comment on this situation, this workshop is the first one I have accepted because of the workshop topic and the venue. The problem emerging here is the type of challenge often undertaken by the National Academies of Science and Engineering. I think that we are at a watershed in public higher education, certainly in California. We are as vulnerable as any institution to temptations, and although I do not see Novartis as predatory, other organizations are.

The analytical capacities of our society are being nibbled away—from the destruction of the Office of Technology Assessment, to the decline of consultants in the legislatures, and the assault on the national laboratories. Many of the institutions of analytical credibility are gone, and others have declined in capacity. And the Novartis situation, unless it is managed well, is not going to help. In our society, there appears to be an increasing hostility—now I speak as a political scientist—to analytical work in general.

I spend a good deal of time at Los Alamos, in part working on matters related to radioactive waste. What I see in important sectors of our society are smart people who discount scientific input out of hand. This is a very difficult situation, and there is a lot to lose, and I do not see the Novartis paradigm or much else presented at this workshop fostering a movement away from this skepticism.

Without more thoughtfulness about how to increase public confidence in us—as a chemical association, as a nuclear community, as a community of thoughtful academics, or as people who labor in the analytical vineyards—we will lose a great deal. I hope that discussions about partnering with industry will have room for such matters because industry has a very important role in countering this loss of confidence. This is a very serious concern. It would be a shame for us to enter the 21st century on the edge of losing public trust and confidence in the enterprise we have given our lives to. I think this is what is at risk.

ATTACHMENT: RESPONSES TO THE CNR–NOVARTIS ARRANGEMENT: AN EXPLOSION OF FACULTY CONCERN

Introduction

The CNR–Novartis arrangement has a number of features that are novel both for the UC Berkeley and for public universities generally in the United States. Some features could result in a substantially increased flow of intellectual capital from the corporation to the research programs of a department's faculty and potentially enrich the economic returns to the campus for research done there. Other features—the relative scale and the relative exclusivity and duration—together seem worrisome to the academic community of UC Berkeley. These concerns are summarized below. They arose from extensive faculty discussions.

Categories of Concern[2]—Planning Process

*In CNR's review of the experiences of other institutions' attempts to establish a relationship between a public land grant university and a major international corporation, what were the lessons learned? What would be the sources of learning for an institutional experimental and observational aspect of a CNR–Novartis evolution?

*To what degree was the agreement signed with the full understanding of the public image aspect for both the college and for the UC Berkeley of the arrangement with Novartis? What are the mechanisms for monitoring and enhancing those images?

I. *Optimize Learning from Novartis and Benefit from Innovation Sponsored in Part by Novartis*

Objective: Assure equalized benefits: to California public, faculty and students, and corporation while maintaining the quality of research (and scholars); assure a breadth of research warranted by a public university.

1. How have funds been allocated in terms of the "public good" research conducted? How is the character and public value of CNR–Novartis-supported research to be judged? On what basis? (For questions concerning public interests, see Section V.)

**How have boundaries been drawn between proprietary and public information/technology work? Have they been maintained?

**To what degree are public UC Berkeley funds being used either to support or to provide infrastructure for research favored by industry? (Are the opportunity costs shouldered by the public sector more than the proposed $25 million of unrestricted funds?) *These funds have not materialized.*

**To what degree are private funds supporting "public research"? Has this been used to argue for increased public funding?

---

[2] Starred questions are from the CNR executive committee; numbered questions are from the academic senate.

**To what degree has the college (as well as Novartis researchers and employed graduate students) gained access to funds, technology, and patented and proprietary technology and information? On what basis has this been assured?

2. To what degree has there been restriction on publication of Novartis-sponsored research (e.g., delay until it is clear that no confidential information is disclosed or delay until one is assured that patents are being secured)? Has the provision for a delay of publication for up to 120 days inhibited the reporting process or seen as a disadvantage in scholarly competition?

3. What have been the advantages for the Novartis arrangement as perceived by researchers, especially graduate students? For the quality of CNR's programs as perceived by CNR faculty?

4. Given the provision on "first right to negotiate . . . for license of discoveries coming from the departmental lab," to what degree has Novartis initiated claims on intellectual property arising from work of colleagues in the department that is NOT supported by Novartis (as well as the work that is supported)?

5. How will patents based on Novartis-supported research be handled? What are the general arrangements for filing patents and copyright agreements?

6. How has the stipulation of the "right to license 30 percent" been worked out? On what basis have calculations associated with the "30 percent of all patents" provision been made?

7. Novartis has the right to license some 30 percent of new findings; how has this worked out in practice? Does this include PMB inventions that are *not* done with Novartis funding?

8. How has the disposition of overhead been worked out and for what functions?

II. *Institutional Development Vis-à-vis Ongoing CNR and Campus Processes and Climate*

Objective: Avoid or minimize institutional distortion regarding undergraduate teaching, graduate education, and faculty governance.

9. How has the agreement affected the general atmosphere of the open academic environment in graduate education? Do students perceive the same type of environment as before the arrangement?

10. To what degree has there been a perceived distortion of the undergraduate teaching program? Has the time devoted to undergraduate teaching changed since the arrangement was put in place?

11. How have the Novartis developments affected the education of graduate students? What advantages are perceived? Have choices for graduate students increased or decreased? Have discussions among graduate students involved in confidential research affected interactions within the graduate student body as a whole? Are graduate students publishing their research within the usual timeframe?

12. To what degree has the PMB maintained its control over the direction, scale, and progress of graduate students (i.e., to monitor and certify the types of thesis research work done by graduate student

research assistants on projects sponsored by Novartis scientists to assure that it would be at least equal in quality to that done with UC faculty; to attract able students to the program; and to assure a "normative pace" and attention toward graduate students as the proportion of students engaged in Novartis "targeted research" grows)?

13. To what degree has this (type of) arrangement (shifting research from a corporate to a university environment) resulted in fewer jobs being available in the marketplace for Ph.D. recipients?

14. To what degree has there been a "set-aside of resources" for unrestricted graduate fellowships, to be allocated by faculty within PMB or CNR who are not in the Novartis agreement?

15. To what degree have Novartis resources affected faculty pay (e.g., sustained summer pay) in introducing inequality of reward?

**To what degree has it been difficult to distinguish clearly between "restricted" and "unrestricted" moneys? Have nonparticipating faculty had access to either?

**To what degree has there been a perceived increase in a culture of "haves" and "have-nots"?

16. To what degree is it perceived that "Novartis scholars and professionals," both departmental and adjunct, have had an affect on faculty governance processes? *Adjunct faculty option dropped.*

**To what degree has freedom of association (e.g., individual faculty working with other companies or no company at all) and/or freedom of research, teaching, and publication been affected by the Novartis arrangement?

**To what degree is it perceived that the arrangement has affected the Division of Agriculture and Natural Resources (DANR) agenda for the CNR; on the impact on CNR's outreach mission?

17. What processes does CNR have to gauge the effects of the Novartis relationship on the college's internal development? What indicators of change are being used?

**To what degree has the ease or difficulty of recruiting faculty and student changed since establishing a relationship with Novartis?

**To what degree has the pattern of teaching, curricular stability, and service performed by faculty work changed since establishing a relationship with Novartis? Has this varied for those faculty supported by Novartis resources?

18. How have the appointments of Novartis scientists as adjunct professors been handled? How many such appointments have there been? *Adjunct faculty option dropped.*

19. To what degree has Novartis-supported work affected the size and number of existing faculty and the potential for faculty salaries paid from non-1990 sources?

20. To what degree have Novartis funds been used (or allowed other funds to be used) to add faculty to the department or college?

21. What are the provisions for exiting or terminating the relationship if it proves unworkable?

    21.1. Specifically, do the rights of either party to negotiate an exit with a single year's notice assume any protections for graduate students with implicit four-year commitments and for staff research associates?

III. *Infrastructure Concerns*

Generic concern: There is likely to be an expansion by 100 people associated with the Novartis contract. This is likely to affect the demand for other resources, including possibly foregoing other future growth elsewhere.

22. Where are Novartis personnel housed on campus space?

23. How have company-confidential, restricted-to-Novartis personnel spaces on campus been managed?

24. To what degree have UC faculty, students, and staff signed nondisclosure agreements?

25. To what degree has there been a change in staff (over that already associated with CNR) on campus?

26. How has the provision of space for PMB been managed? Who funded this? Did it come at the expense of others on campus? To what degree has the availability of Novartis funds affected interactions with federal agencies on capital expenditure matters?

IV. *Governance of Institutional Evolution*

Generic concern 1: The degree of exclusivity is novel and a matter of continuing concern. This calls for a structure to assure sustained thoughtful attention.

Generic concern 2: It is naive to assume that PMB can walk away from such a contract with Novartis should the research agenda be somehow "bent" away from the research of the university and toward more applied activities. One cannot easily shut down a relationship of this nature, and thus successful governance is critical.

27. Given the present governance structure, to what degree is this structure seen as being composed of people who will benefit?

28. Have Novartis representatives played a major role in departmental governance or resource allocations? To what effect?

29. How has the distinction between "unrestricted" research funds, paid out of the annual Novartis grants, and specifically targeted research projects been implemented by the research review committee? To what degree have Novartis representatives on the two boards been influential in decisions about nonrestricted, non-Novartis-related research?

30. What provisions have been made by the college (department) to prepare for the analysis of feedback from such study and to effect remediation if necessary?

31. Has there been a vehicle to provide a voice for the CNR faculty and academic senate in matters of reviewing the institutional outcomes of the CNR–Novartis evolution?

32. To what degree have the interests of the senate been represented via the current arrangement to university representatives (UC Berkeley, vice chancellor for research, and the University Office of the President and Division of Agriculture Natural Resources vice president)?

V. *Optimize Relationships with a Wide Variety of California Stakeholders*

Generic concern: Expand range of public-good research to include awareness of public and social impact and potential mitigation as well as benefits.

33. How have plans worked out for developing a research capacity, not only to develop new agriculture products and methods, but to provide insights into the consequences of their widespread deployment?

34. How has the Novartis arrangement informed research relationships with the California commodity groups?

35. The principles outlined include a provision of no oversight by the industrial partner. But both the proposed research allocation committee and the policy board include such individuals. How has this worked in practice?

36. What processes have been put in place to involve stakeholders in receiving information about Novartis-based developments and seeking advice for research directions? How has this been received?

VI. *Experimental Nature of This Development*

Objective: To provide the basis in advance for a rigorous experimental design and resources for its timely execution.

37. What processes have been put in place to assure that all parties understand the experimental nature of this agreement and the importance of acute feedback and analysis of "lessons learned"?

38. How has timely feedback been provided to oversight group(s) for midcourse corrections if appropriate? (Who has been involved and on what grounds?)

DISCUSSION

Hank Whalen, PQ Corporation: How did you get Novartis to sit still while these negotiations went on? And how long has this been going on?

Todd La Porte: I don't know the history of this in a careful way. It didn't take very long, possibly because there were several people at Novartis who realized, through whatever series of experiences they had had in the past, that they would to be locked out altogether if they didn't pay attention. If I recall correctly, Novartis and UC Berkeley were in conversation for about 18 months.

David Schetter, University of California, Irvine: Does the agreement provide for the university to have a role in the selection of the projects that are being carried out?

Todd La Porte: Yes. That is the role of the research committee. The disbursements on the list of projects range from $200,000 to $75,000 per faculty member based on the selection committee's sense of quality.

David Schetter: I would say that there is more to this that is positive. First of all, there is the flexibility of the research mission, which is not uncommon for Novartis. Also, there was a $25 million building for UC Berkeley.

Todd La Porte: Not exactly; it became a $3 million facility. The $25 million somehow disappeared. I am familiar with the facts.

David Schetter: I thought it was a higher price.

Todd La Porte: I agree. For 30 faculty and $3,300, its not that much per person. In fact, one of my colleagues in the Division Council, which is the body that worries about these things, said it should have been three times per person.

David Schetter: It puts enormous pressure on the other universities that are not doing these "creative deals." The other part is that, whereas they can access only 30 percent of the licenses, they want to look at everything. And if there is any joint invention, what good is it? Essentially they are going to capture over 60 or 70 percent of actual licensed information.

Todd La Porte: This wouldn't surprise me. We already came up with that hypothesis, and it is one of the questions in the attachment. But don't misunderstand the tone of this conversation. I sounded like I was being negative, but I'm not. As a faculty member, I'm equivocal because we don't know what the requirements are for engaging with industry this way and doing it gracefully. If we can learn whether it can be done gracefully or not, then many of the things that we here are supposing can be seen as hypotheses rather than conclusions or predictions. As a biologist, for example, or in my case as an organization studies person, we don't know enough about the details of these agreements. We are told what the organizational outcome will be. But my response is, "Wait a minute. You can't tell me that your confidence in the goodness of your heart and your colleagues will result in that kind of outcome with a high degree of certainty." It might and it might not. It's indeterminate. And that's why the "organizational following" is necessary.

Joseph Gordon, IBM: This is a single, relatively large grant or contract. Is that the same as a number of small ones that add up to the same amount of money? Are there the same concerns?

Todd La Porte: I made the allusion that we know how to handle a number or consortia of sponsors and a group of faculty. We do this a lot, and many of you do too. Why do we say that? Because sponsors check each other. Essentially none of them are large enough to have the potential for untoward influence if that were to happen. The fewer contracts you have, the more money involved and the longer the relationship is, then the more you have to wonder about it in terms of institutional care. And that is why we are in the middle between the arrangements we do quite often and the national laboratories. The Novartis deal is in the middle, and we do not know yet how to do this type of agreement gracefully.

The question is whether we can figure out a way of doing this gracefully and not distorting the meaning of public higher education in the 21st century. That is the basis for the concern because higher education is what is at stake here. As long as our public and our public legislative bodies are either distracted or maliciously disinvesting in analytical capacity, we somehow, as a university and as an analytical community, have to figure out ways of maintaining our analytical capability, and particularly the capacity for freely posing questions. It's the question-posing part that is as important as anything else, and that is hard enough in any environment to maintain.

Claude Meares, University of California, Davis: As I'm sure you remember, something related to this happened at Scripps about five years ago. How do you see that as relating to this?

Todd La Porte: Do you know enough about it to tell our colleagues here what the Scripps situation was?

Claude Meares: It was with Sandoz, another large Swiss corporation. This was one of the director's entrepreneurial activities, which at that time attracted the attention of Congress because it involved a share of the inventions, similar to the UC Berkeley and Novartis agreement, that might have been supported by federal funds and that was renegotiated after considerable rancor. It seems to me that something must have been learned that impacts the UC Berkeley and Novartis case.

Todd La Porte: One of the questions listed in the attachment was to what degree the planners at Berkeley have taken into account the history of the UC and these types of agreements in the past. You can conclude that the UC Berkeley people were aware of some past events and were hoping it wouldn't happen to them. But it was certainly within the consciousness of most of the people who were close enough to that research community to know the full story.

And there's more. At Irvine there has been an unfortunate relationship with a Japanese firm and also at UC San Francisco. So there's a lot of concern here that has surfaced. I have no reason to believe that other universities are not having the same experiences.

Francis Via, General Electric: I have a couple of concerns with the basic premise before you began the discussion of the Novartis agreement. I'm relating myself more with the discussion that occurred on campus between the "haves" and the "have-nots." Being a chemist and being on the Chemical Sciences Roundtable, I'm starting to feel like the "have-nots" associated with the type of agreement you are able to work out at the university. And in trying to picture what would happen if such a proposal were made to a general chemical company on general chemical technology, the consternation is with the initial one, that unless the university, a world-class leading university, teams up with a corporation at this stage to gain knowledge from the company, the university would be ineffective in educating its students. We have seen the content of teaching students to solve problems, and basic science seemed to work for a 100 years or so. I'm wondering if that premise is fractious?

Todd La Porte: No. I'm glad you raised the question because that is one other aspect of this "observational following." We are going to be asking two questions. First, do the design principles return an outcome that was expected? We're going to try to test the principles. Second, if the outcome is as expected, is this the only way to provide access to the proprietary data that in this particular case you can't get anywhere else? I don't understand the specifics of the agricultural genomic database. What my understanding of it is, in talking to the faculty about why this is a problem, is that because it's so fundamental, that if certain companies or certain bodies hold pieces of information exclusively, the department cannot use it with regard to the education of their students. It's not the process part of education that you were referring to. It's the substance aspect of it. Let me tell you a story. UC Berkeley had several instances in which students had gone through a dissertation project, finished the dissertation, and taken it out to the world. But the corporation said, "Oh, we already knew that, and we knew it two years ago, but we weren't telling anybody we knew that." Now, if you are a graduate professor with students working as hard as they do, to program that kind of outcome, it's just plain irresponsible. Those lives are too precious and those skills are too rare to treat them as temps or to treat them as dispensable. If you can avoid that, you should. That is part of what's driving the character of the Novartis agreement, and that is what's novel from what is happening on our campus. This is happening, it turns out, in the computer sciences too. We are beginning to ask the question: Where else is this happening?

Andrew Kaldor, Exxon: It's happening all over.

Todd La Porte: If it is happening all over, then it is our responsibility as academics to ask how we can prevent damage from this in terms of the graduate students and as a public institution. If you are a private institution, it's not a problem. But if you want to align yourself with a special interest, do that. There's no public prohibition from this in their case, but in our case there is. It's not only illegal, it's certainly the way my colleagues feel about their responsibilities to the rest of the society as well as to their students.

I want you to tell me that I'm wrong, but I'm not hearing anyone say that I am wrong. I'm testing this out on you because you should know if the dynamics that I am describing exist because it's a small sample and because I'm not involved in the science. And I'm not hearing from you or my colleagues, who are in the Novartis agreement at Berkeley, that we are wrong about this, and that is scary.

Michael Chartock, Lawrence Berkeley National Laboratory: I find many of your comments and arguments very interesting and compelling, but just because I am silent does not mean that I agree with all of them. Actually, I just have one small qualification. You didn't say that the Lawrence Livermore National Laboratory (LBNL) necessarily conducted classified research, and you say that the University of California had a good experience with outside research conducted at the national laboratories. But the LBNL does not conduct classified research.

Todd La Porte: That is quite true. I appreciate that, but it's instructive why that is the case. Some time ago we decided that we couldn't have a good relationship with LBNL while it was conducting classified research. So part of the response to doing classified research is to figure out a way for them to exist without having to do it. So we have given it to Livermore and Los Alamos, which have a different kind of relationship with us as a university and as a campus.

Michael Chartock: The handling of classified information and where you do it is very complex. Long before 1970 when the LBNL separated from the Lawrence Livermore National Laboratory, the Department of Energy had taken much of the classified research out of the laboratory. I think the department and the federal government realized that national labs were probably not the best location for doing classified work.

Christopher Hill, George Mason University: I'd like to consider the comment about whether the state of industrial practice is in some sense so far ahead of what the universities know that we can no longer effectively teach what we know in anticipation of what will happen in industry in the future. Embedded in this comment is the implication that in the good old days we faculty members were always better informed and ahead of the state of the art of colleagues in industry. To the contrary, it seems to me that industry practice has nearly always been ahead of academic teaching.

Todd La Porte: That's not the point here. Your point is also true, but in the past if we wanted to find out we could. Now we can't. We are sending our graduate students out to essentially enlarge their knowledge from outsiders. Right now this has declined in the view of the colleagues in that department. They say that this is the situation they confront. Their belief in this was strong enough to be troubling and they have had experience doing it. I have a lot of cunning colleagues at UC Berkeley in this matter. It seems clear that we need to learn some things that we don't know and that are not generally known about this type of relationship.

Debonny Shoaf, Idaho National Engineering and Environmental Laboratory: A couple of observations: You commented about the inaccessibility of some kinds of information because information is closely held, and that there exist proprietary constraints with respect to sharing information. As far as I know, sequestering information has existed as long as there has been intellectual exchange. I agree with you that it is important to question, to explore, and to contest the constraints on free flow of information; however, I believe that there are often good reasons and strong drivers for nondisclosure, and any attempts to change that would have little impact.

The other observation is one that might be a little bit flip, but as you were referring to organizational models, you referred to the "predatory" nature of the industrial enterprise. I would refer you to an organizational model presented by microbial communities in nature. Although a population may be subject to predators, a good predator never completely destroys its host population.

Todd La Porte: Then you want to make sure that you are involved with a good predator.

7

What Industry Seeks from Federal Laboratories, and Examples of Collaborations

Henry T. Kohlbrand
Dow Chemical Company

We work in a rapidly changing business environment. Companies that were formed when regional economies controlled market forces are now competing in a global business environment. This has resulted in industry consolidation and changes in many business practices. This is especially true in mature industries, such as the chemical industry.

The Dow Chemical Company, formed in 1897, has had a long history of developing technology within the company as well as working with universities, national laboratories, and institutes external to the company. The Dow external technology organization today coordinates over 150 external research programs and a significantly larger number of external research agreements as well as participating in a number of business development and growth initiatives. This paper focuses on how these collaborations bring value to a multinational company, key factors that impact the success or failure of these activities, and challenges for the future. In addition, several examples of collaborations between Dow and federal laboratories are reviewed.

THE CHANGING NATURE OF BUSINESS

The rapidly changing business environment in which we work requires us to continuously improve the processes that govern our activities. The move from regional to global economies has resulted in significant industry consolidation and restructuring. This has impacted both the mechanism and the role of technology development within large multinational companies. Where manufactured products were once the focus of activity, the management and exploitation of intellectual assets, once considered only an enabler to manufacturing, have now become much more economically important. In many cases, information and services have become the product.

These changes in business structure have had a significant impact on how external collaborations are managed. We are required to move quickly while maintaining a high degree of accuracy. The degree of technical understanding needs to be balanced against project timing and customer needs. Demands on product quality and performance increase while pressure to reduce prices continues. Competitive assessment has become more important, especially considering competitors that are new to the marketplace. Joint ventures and acquisitions have become a much more common way of doing business; and purchased technology, rather than technology developed within an organization, is playing a more important role. On top of all of these changes, reengineering and restructuring have resulted in fewer employees in large companies. External collaborations must be structured to use people's time as effectively as possible.

External partnerships are now more important than ever. Understanding how to form alliances and generate value from these alliances is a critical factor in meeting the challenges of the next decade. The definition of ownership of intellectual property resulting from an external collaboration is more important than ever. In the past, the intellectual property resulting from collaboration was directed primarily toward the sponsoring company. Today it is quite common for this property to be directed toward a joint venture or later licensed or sold to another company. Often, three or more companies are involved in the use of the intellectual property. The agreements associated with ownership of this property must allow increasing flexibility in how it will be commercialized.

OVERVIEW OF DOW

The Dow Chemical Company is a global science- and technology-based company that develops and manufactures a portfolio of chemicals, plastics, and agricultural products and services for customers in 168 countries around the world. With annual sales of more than $18 billion, Dow conducts its operations through 14 global businesses employing 39,000 people. The company has 123 manufacturing sites in 32 countries and supplies more than 3,500 products. In 1998, Dow divided its 14 businesses into 7 segments.

Table 7.1 shows the diverse nature of the businesses. Each of these operating segments has very different needs for different types of external collaborations. In addition to business diversity, geographic diversity also plays a role. In 1998, 40 percent of Dow's sales were from the United States, 35 percent from Europe, and 25 percent from the rest of the world.

R&D at Dow has been significantly redirected over the past few years. More effort is being placed on growth. Dow's R&D investment in maintaining current businesses is decreasing in favor of R&D focused on value growth. The R&D budget at Dow has gone from $785 million in 1997 to $807 million in 1998 to a projected $870 million in 1999. Much of this additional investment is directed toward value growth, with a significant commitment to biotechnology. We are also increasing our attention on opportunities in biomaterials, coatings and adhesives, electronic materials, and special polymer processing.

The changes in business structure at Dow have impacted the nature of our external collaborations. Prior to 1994, Dow's organization was a matrix of functions, businesses, and geographies. In 1994, a major reorganization took place. Each of the 14 businesses formed contained its own commercial, manufacturing, and R&D organization. Shared or leveraged organizations such as environmental, health, and safety; controllers; supply chain; and corporate R&D continued to exist, but their focus was directed toward the success of the 14 business units. In 1998, corporate R&D reorganized into eight capabilities. These capabilities are very aware of the business needs and are responsible for leveraging across businesses as well as developing their capability for the future.

All of the technology collaborations sponsored by Dow are supported either by a business or by a corporate R&D capability. Eighty percent of the external projects are supported by one of the 14 businesses and 20 percent of the external collaborations are supported by corporate R&D capabilities. When we establish an external technology partnership, the sponsoring business or capability will set the scope and timing of the projects. When working with Dow, external collaborators need to work with and be aware of the needs of 22 internal organizational units—the 14 businesses and the 8 corporate R&D capabilities. We have an external technology organization that addresses this interface (see Box 7.1).

TABLE 7.1   Operating Segments and Sales

Segment	Businesses	1998 Sales (%)
Chemicals	Basic chemical products	13
Performance chemicals	Specialty chemicals Emulsion polymers	14
Plastics	Polyethylene Polystyrene Polypropylene Insite technology™	20
Performance plastics	Adhesives, sealants, and coatings Engineering plastics Epoxy products and intermediates Fabricated products Polyurethanes	28
Agricultural products	Crop production; weed, insect, and plant disease management; industrial and commercial pest management	13
Hydrocarbons and energy	Fuels and crude-oil-based raw materials	8
New businesses/unallocated	Technology licensing; advanced materials for electronics; industrial biotechnology; venture capital	4

BOX 7.1
Corporate R&D Capabilities

- Analytical sciences
- Biotechnology
- Catalysis
- Computing, modeling, and information sciences
- Engineering sciences and market development
- External Technology and Intellectual Asset Management
- Materials Engineering Center
- Materials research and synthesis

EXTERNAL TECHNOLOGY PARTNERSHIPS

External Technology and Intellectual Asset Management is one of the eight corporate R&D capabilities. This organization manages over 150 external agreements with over 100 different institutions. About 60 percent of these collaborations are in the United States, 20 percent in Europe, 7 percent in Asia, 7 percent in the former Soviet Union (FSU), and 6 percent in other parts of the world. The number of agreements in Europe, Asia, and the FSU is growing faster than are those in the United States. In addition, it manages 14 government contracts with a face value of $60 million.

There are many business reasons for collaborating externally for technology development and acquisition. Our external collaborations bring speed, additional expertise, and broader scope to our projects. Often, we can share the risk of working in a new area by involving partners outside of our company. The objective of our external collaborations is to complement Dow's capabilities and to fill business needs through external partnerships and to bring projects to commercialization more quickly and more cost effectively.

The major segments of this capability are shown in Box 7.2. Each of these segments represents an area of activity for the corporate capability. Cooperative research involves collaboration with universities and institutes; government–laboratory collaboration involves joint projects between government laboratories in the United States, Europe, and Asia and Dow Chemical Company. Government contract R&D includes the development of new contracts as well as the management of the $60 million in contracts currently under way. Although most of these are based in the United States, there are some in Europe and Asia. We also work collaboratively with a number of government agencies throughout the world. In the United States, examples of such projects include Vision 2020 (with the U.S. Department of Energy [DOE]), Green Chemistry (with the Environmental Protection Agency [EPA]), and the Partnership for Advancing Technology into the Housing Industry (with the U.S. Department of Housing and Urban Development [HUD]).

Our effort in geographic development includes the building of technology networks in areas of the world where Dow has a limited but growing presence in the commercial and manufacturing arenas. Currently we have efforts under way in the FSU, Israel, China, India, Korea, Australia, and Taiwan. These networks are in addition to the strong external networks in place in North America, Europe, and Japan.

BOX 7.2
Segments of External Technology and Intellectual Asset Management

- Cooperative research with universities and institutes
- Government laboratory collaboration
- Government contract R&D
- Government agency interactions
- Geographic development
- Business development and growth
- Technical advisory boards
- Intellectual Asset Management Technology Center

The External Technology and Intellectual Asset Management capability also provides leadership and coordination for a series of technical advisory boards. These boards invite experts from outside of Dow to provide our businesses and capabilities with an external peer review of our activities as well as the opportunity to participate in collaborative projects.

The final segment of our capability is the Intellectual Asset Management Technology Center. Although the management of intellectual assets is done within each business, this technology center provides best practices for all Dow businesses in intellectual asset management, as well as providing training and career management for those involved in this activity.

The process for sponsoring external projects within Dow has changed significantly during the past ten years. At one time, about one-third of the funding of external projects came directly from corporate funding. There were virtually no projects in countries such as the FSU, China, Israel, etc. Although done with links to businesses and R&D groups, the accountability was often unclear. Today, all collaborations are sponsored by a business or corporate R&D capability, and accountability is very clear and there is clear growth in agreements in our new geographic networks.

There are several elements common to successful external partnerships. First, it is necessary to be proactive in using external partnerships. Next, it is critical to have a vision for the successful outcome of the project. This vision may be slightly different for each partner, but success should be defined from the start. A part of defining success is to outline a project where all of the partners can "win." Assembling a successful collaboration, therefore, involves each partner understanding what success means to the other. Definitions of success are different for a company, university, and federal laboratory, but it is critical that these differences are clearly understood. Continuous communication during the project also enhances the likelihood for success. Finally, as a project goes forward, it is important to keep focused on how we can continuously improve the way that we do things. If these elements are observed, we have found that our collaborations are much more likely to succeed than if we just let them happen. One of the functions of our corporate R&D capability is to assure that these elements are a part of each new collaboration.

FEDERAL LABORATORY INTERACTIONS

Over the past 20 years, Dow has had some sort of collaboration with most of the federal laboratories. During the past five years, we have had projects with the following laboratories:

- Argonne National Laboratory,
- Armstrong Laboratory (Air Force),
- Brookhaven National Laboratory,
- Lawrence Berkeley National Laboratory,
- Los Alamos National Laboratory,
- Naval Research Laboratory,
- National Institute of Standards and Technology,
- Oak Ridge National Laboratory,
- Pacific Northwest National Laboratory, and
- Sandia National Laboratories.

There were approximately 50 agreements with these laboratories during the past five years, of which 14 are active today. The following examples represent a cross section of these agreements with a few comments on success factors for both successful and unsuccessful projects.

Argonne National Laboratory

Dow and Argonne have had a number of successful interactions over the years. One successful collaboration involved a Dow scientist who spent six months at the laboratory. He was involved in a series of technology areas and eventually focused on some coating technology of interest to Dow. This coating technology is undergoing commercial evaluation and has a good chance of being incorporated into a new product area.

Dow and Argonne were also involved in the joint development of technology related to fuel cells. Although the technology associated with the project was successful, Dow made a business decision to exit the fuel cell business and so the project was never completed.

Los Alamos National Laboratory

For several years, a Dow scientist worked at Los Alamos. Through his interactions, a number of successful projects were initiated. This scientist was chosen not only for his scientific capabilities, but also his leadership and management capabilities and his ability to develop external collaborations. By being on site for several years, he built strong relationships as well as a thorough understanding of the capabilities of both Los Alamos and Sandia.

One of the projects identified by this scientist was not successful. It involved chemical synthesis. After six to nine months, the person sponsoring the project within Dow changed jobs and the project lost its champion. The person who took his place never developed the same vision for the project and it was never completed.

Oak Ridge National Laboratory

We have had a number of successful projects with Oak Ridge in both ceramics and analytical development. These projects shared four common characteristics: They met a defined business need for Dow and met the mission of the laboratory, each had a champion in Dow and at Oak Ridge, each had defined deliverables, and there was a win–win proposition for both Oak Ridge and Dow.

Pacific Northwest National Laboratory

Our recent collaborations with Pacific Northwest have focused on information technology. This project is just beginning but exemplifies all of our current success factors. We have defined our needs and believe that there is a win–win proposition for the laboratory and Dow. Project champions in both organizations are in place and communication is continuous. This is clearly an area of technology that will not be a core technology to Dow, so it is an area in which we depend on external sources for both expertise and execution.

Sandia National Laboratories

As with Argonne, we had a project related to fuel cell technology with Sandia. Although the technology associated with the project was successful, Dow made a business decision to exit the fuel cell business and so the project was never completed. We have also had a very successful project on materials modeling. Here, Dow has a business focus, but the basic technology being developed can be used on more than one business application. There is a clear understanding of success for the partners and strong project champions in place.

MULTIPHASE FLUID DYNAMICS CONSORTIUM

Another way for companies, universities, and federal laboratories to work together is through the use of a consortium. Once a key area of interest has been identified, the consortium can be organized and the participants selected for the strengths of the capabilities that they bring to the partnership. The Multiphase Fluid Dynamics Consortium is a group of laboratories, industries, universities, and government agencies partnering to perform fundamental multiphase fluid dynamics research. In this case, it was decided to focus on solid-gas processing.

Gas-solid and gas-solid-liquid flows are identified in *Technology Vision 2020: The Chemical Industry*[1] as critical to developing advanced chemical reactors and separations. Solids handling is a critical technology for chemical manufacture. For companies such as DuPont and Dow Chemical Company, more than 50 percent of the products sold are in a solid/particulate form. This consortium was formed to develop technology to accurately model gas-solid transport in industrial applications.

As with other worthwhile efforts, a successful consortium activity does not develop overnight. Prior to 1995, a series of collaborations were taking place across the chemical industry. Dow, DuPont, Chevron, and Exxon all had projects under way at a number of universities and federal laboratories. The CRAY Chemical Process Computational Fluid Dynamics (CFD) Users Group, The Reactive Multiphase Flow Workshop held at Los Alamos in May 1995, the First Engineering Foundation Conference on CFD in Chemical Reaction Engineering held in San Diego in October 1996, and the DOE Office of Industrial Technologies (OIT) Roadmap Workshop on CFD held at the University of Maryland in October 1996 all demonstrated that a significant collaborative effort was required to address the opportunities identified in the Vision 2020 document.

In 1998 the consortium was formed. The OIT supported two related projects in the area of CFD, which form the keystones of the consortium. DOE funding was awarded at a level of $1 million per year. The industrial leaders (Chevron, Dow Chemical, Dow Corning, DuPont, Exxon, and Siemens Westinghouse Power, with computer maker Silicon Graphics and engineering software companies AEA Technology and Fluent) provide leadership and cost sharing. DOE funding supports work at Lawrence Berkeley, Los Alamos, Oak Ridge, Pacific Northwest, and Sandia National laboratories and the Federal Energy Technology Center. The consortium also provides funding for related research projects at Clarkson University, Illinois Institute of Technology, Princeton University, Purdue University, and Washington University-St. Louis.

Each partner in the consortium contributes through their strengths and knowledge in a variety of areas. Their knowledge in defining needs, experimentation, equipment design and construction, and modeling as well as application experience all come together to guide the projects. A clear definition of industry needs, project milestones and timetables, and desired outcomes help to measure progress and direct activities. A path to share the results through a series of meetings, publications, and presentations is also in place. Also, strong project champions in each company, university, and federal laboratory play a key role in making sure that the project targets are met. Tyler Thompson of Dow and Brian Volintine of OIT played key oversight roles in organizing the consortium.

---

[1] American Chemical Society, American Institute of Chemical Engineers, Chemical Manufacturers Association, Council for Chemical Research, and Synthetic Organic Chemical Manufacturers Association, *Technology Vision 2020: Report of the U.S. Chemical Industry* (Washington, D.C.: American Chemical Society, 1996).

Several key factors have contributed to the success of this effort. At the beginning, a lot of effort went into establishing a shared vision among all of the participants. The team leading the effort is both self-organized and self-directed. Each organization participating is committed to a portion of the project, and the industry is ready to implement the results as they become available. It is also important to note that the technology resulting from these projects results in precompetitive tool development rather than invention and discovery that might be considered proprietary by the participants.

SELECTING PARTNERS

All institutions are proud of their capabilities. This is true of industry, federal laboratories, universities, and institutes. Each institution has a few capabilities that truly differentiate them from others. When we start to form a relationship with a new institution, we try to identify these differentiated areas to begin projects in these areas. It is very important that federal laboratories, universities, and institutes clearly communicate what they consider their strengths to be and make it easier for collaborators to find their way through "organizational mazes."

WHAT WE HAVE LEARNED

In our work in developing external technology partnerships, we have found a series of elements that need to be present in order to have a successful project. These are listed in Box 7.3.

A project that is well defined at the beginning has a much better chance of success than one in which the definition is vague. An understanding of the company's needs as well as the laboratory's capabilities sets the foundation for defining the project. In our case, each project needs to fit a defined need by one of our business units or one of our corporate capabilities. A project without such a connection has little chance of being funded. Early in the project, the agreement on results, how to measure them, and intellectual property ownership contributes to the focus of the principal investigators and leaves less room for misunderstanding between partners as the project progresses. Effective communication during the project helps to keep the project on track and allows the partners to agree on necessary changes in plans. Finally, successful projects need to deliver results.

BOX 7.3
Elements for Successful Projects

- Define the project well at the beginning
- Make sure that the federal laboratory develops an understanding of Dow's needs
- Make sure that Dow has an understanding of the laboratory's capabilities
- Link the proposal to a specific Dow business or technology need
- Agree on results, measures, and ownership of developments early in the project
- Develop a project that benefits all partners
- Communicate often
- Update scope and deliverables
- Update definitions of success and satisfaction periodically
- Deliver success

Most successful interactions involved Dow scientists and external technology staff working with federal laboratories to clearly understand their capabilities before any projects were defined. Almost no successful interactions occurred as a result of laboratory outreach visits (although once a relationship is established, visits from laboratory staff and scientists help to build understanding among company scientists). Each successful project had a project champion inside of Dow and a project champion at the federal laboratory. Where laboratories made a special effort to identify their areas of strength, interactions seemed to proceed more quickly and more effectively. Some projects had unique challenges because some of the company participants were outside of the United States. The need to involve international teams is increasing as the global presence of companies continues to grow.

In building partnerships, it is important to remember that there are four partners in each company–federal laboratory collaboration:

- Federal government,
- Federal laboratory,
- Sponsoring business or research capability in the company, and
- The company.

Many times an agreement takes additional time to finalize because one of the four partners is not taken into account. It is important to recognize that the needs of the federal government and federal laboratory do not always align, and, similarly, the needs of the company and sponsoring business unit many not always align. Reconciling these issues early in developing a relationship can avoid misunderstandings later in the project.

Flexibility is increasingly important. In the examples of Dow and federal laboratory interactions, it was pointed out that a project can be technically successful, but may not be completed because the business needs of the company change. Needs of both organizations may change over the life of the collaboration. Continuous communication helps to manage these situations more effectively.

The exchange of people facilitates knowledge and technology transfer. It would be ideal to have company scientists present at each federal laboratory, university, and institute. When company scientists have been stationed at a federal laboratory, the number of quality collaborations has greatly increased. Similarly, when federal laboratory scientists have worked at an industrial facility, the number of quality interactions has also increased. Unfortunately, this is not practical. The number of people required and costs associated with this effort would be prohibitive. This is possible in a few, very select cases, but alternatives to this mechanism for knowledge and technology transfer must be found.

Any industrial–federal laboratory collaboration must be cost effective and cost competitive with alternatives. This is an important aspect of defining success at the beginning of an interaction. Part of this discussion involves the ownership of intellectual property resulting from the joint activity. This tends to be easy for precompetitive projects, but can be increasingly difficult for projects that are closer to market implementation. In some cases, ownership of patents resulting from joint efforts may be governed by legislation and may restrict the scope of the project.

Finally, people make the difference. With the correct people involved and teamwork established, the likelihood of project success can be increased.

FUTURE OPPORTUNITIES

There are always opportunities to improve the ways that we work together. It is often difficult for the federal laboratories to know where they should be developing expertise for the future. Better communication by industry of its needs is an area in which we need more focus. Efforts such as Vision 2020 have begun to collect some of this information, but in general, more specifics are needed.

We should start with an understanding of how to make collaboration a win–win situation. Key elements need to be defined from the beginning. Understanding how sensitive or proprietary the project area is to the company, or how important it is for the federal laboratory to publicize the collaboration, can shape the agreement more effectively. In one interaction we had with a federal laboratory, the premature publication of the collaboration by the laboratory made the project impossible to complete.

Intellectual property issues need to be addressed more effectively. In the past, collaborations typically involved a company and a federal laboratory. Simple assignments of the patents or provisions for the company to license the technology were adequate. The DOE cooperative research and development agreement (CRADA) "Subject Invention" definition can confuse intellectual property ownership. Today, the collaborations may involve several partners, the results may be used by joint ventures, or even more-creative business structures may need access to the intellectual property developed by the collaboration. Increased flexibility on how this property is managed is a key issue for the future.

When developing relationships with federal laboratories, it is sometimes desirable for the company to invite federal laboratory employees to act as consultants. For scientists who are employees of the U.S. government, U.S. law declares their inventions to be U.S. property (42 U.S.C., Sec. 5908) and makes it difficult for them to work as consultants. Some privately operated federal laboratories apply these same rules to their employees. Others operate on a different set of rules. This is an area in which improvements and clarity would be welcomed.

CONCLUSIONS

Industry looks for excellence, partnership, cost effectiveness, and versatility in external collaborations. The federal laboratories are a great resource for joint technology development. A better understanding of the laboratories' strengths by industry will aid in the most effective utilization of their capabilities. Key success factors include defining a project in which all partners' needs are met, having a project champion in each organization to facilitate progress, communicating continuously, and defining flexibility into the agreement so that it can change if necessary. A moderate effort to broaden options for ownership of intellectual property and involvement of federal laboratory employees as consultants could facilitate additional collaborations with industry.

DISCUSSION

Bruce Harrer, Pacific Northwest National Laboratory: Most of the employees of the national laboratories are not federal employees, so I do not understand the issue over intellectual property.

Henry Kohlbrand: I cannot answer that from a legal point of view. Intellectual property covers federal employees; it covers apparently a broader spectrum than that, according to our lawyers. This was a problem at Sandia. Sandia lawyers said that they couldn't sign this kind of a secrecy agreement because we're covered under it. So I can't answer the legal question of that, but I can tell you that this question has been raised by the legal staffs of several of the federal labs that do not have federal employees.

William Millman, U.S. Department of Energy: The difference between federal laboratories and national laboratories is that federal laboratories have federal employees. They are government owned and government operated. The DOE's large multipurpose national laboratories are government owned, but contractor operated. The result is that laboratory employees are employees of the contractor, such as the University of California or Lockheed Martin.

Henry Kohlbrand: I cannot adequately answer the question. It's a question that we should highlight from this discussion. I know that the lawyers are worried about it.

Ashok Dhingra, Dupont: Would you comment on Dow's collaborations in Asia Pacific, specifically how you deal with the communication and proprietary issues?

Henry Kohlbrand: We have had some good luck. We've had an office of science and technology in Japan for a long time, and we found that working in Japan is a little different from a cultural point of view. Japanese agreements are wonderfully short. The Japanese say that we are going to work for mutual benefit and that the details can be worked out later. If you are not comfortable signing something like that, then you're in trouble.

We have several interactions going on in China and India. Because of the intellectual property issue, we have tended to stay away from areas of technology that are what I call sensitive within the company right now. So we might be working with the National Chemical Laboratory in Pune or we might be working with the Indian Institute of Chemical Technology on some specific projects of developing some new chemistry or some new catalysts. In China we are doing a lot of synthetic organic chemistry with some of the Chinese institutes, and we are doing some engineering work with one of the large universities in China.

So we try and recognize that there are differences in the way proprietary information is handled within these countries, so we don't bring a sensitive project to those countries, but are effectively working with them on other things.

Christopher Hill, George Mason University: My comment is on the definition of a federal lab and a national lab. Some of the federal labs that belong to DOE have been designated "national labs," but all of them are federal labs.

As the structure and decision making for collaborations within Dow have changed in recent years, has there been a concomitant change in either the functional level or responsibility of the individual within Dow who has the sign-off authority on engaging in a collaboration? Has that moved up or down in the organization or away from the R&D management? Where does it reside?

Henry Kohlbrand: Depending again on the size of the agreement, I would say, in general, that the signing authority may have moved down a notch because we have gone from roughly 13 levels of management down to about 5. We cut out extremely large numbers of management layers. But for large programs, we need to get business approval. So it is not just the R&D community that approves collaborations.

Sometimes we have to help the business people understand why it's necessary to be spending this kind of money outside the company and what we expect to get for it. In some cases if it's money that's being given as a nonrestricted grant to avoid overhead and to do some other things with the university, then the business people have some struggles with that. So we have to do additional communication and additional work within the organization. For most of the research projects that are taking place, approximately 80 to 90 percent of them, they are still handled within the R&D umbrella, and we don't have those types of problems.

Hank Whalen, PQ Corporation: Are CRADAs increasing or decreasing with Dow?

Henry Kohlbrand: Right now we may have one or two more CRADAs this year than we had a couple years ago. I don't know if that is a substantial increase. Again, CRADAs are a good way to interact with a laboratory as long as you work together on how the outcome of the research is going to be done. We are seeing more involvement in the Advanced Technology Program (ATP). We are more interested in the company and ATP-type programs than we have been in the past because we can get closer to products with ATP projects than we can with CRADAs. So I would say that it is flat right now. I am unsure if there is a reason to drive it up and down at this point.

James Seiber, U.S. Department of Agriculture: You gave a number of examples of cooperation with national laboratories. What are your experiences with other federal organizations, such as the Department of Agriculture, the National Institute of Standards and Technology (NIST) and the EPA? And maybe more importantly, how do such organizations make the first contact within your company?

Henry Kohlbrand: I am unaware of any interaction that Dow has had with the Department of Agriculture. That doesn't mean that there aren't any because Dow is a big company. In terms of the other organizations, we have worked extensively over the years with the Department of Transportation because we transport a lot of stuff. We have also worked with the Coast Guard and the Department of Commerce. We have a lot of interactions with NIST because NIST is a marvelous place to do some types of work, so that is a continuing effort. We have a person who goes out two or three times a month to foster some of these interactions. When working with the EPA, we have had both good and bad experiences, as you might well imagine, because there are a couple different parts to the EPA. But I had some good interactions with the technical side of EPA, when we were collaborating on incineration when I was working in that area about ten years ago. So Dow has a lot of activity going on with the agencies, and we have two people in Washington, D.C., to facilitate these activities.

8

Partnering at the National Laboratories: Catalysis as a Case Study

Nancy B. Jackson
Sandia National Laboratories

THE NATIONAL LABORATORIES

The role of the national laboratories—particularly the defense program laboratories—since the end of the Cold War has been a topic of continuing debate. The relationship of the national laboratories to industry spurred debate ranging from designating the labs as "instrumental to maintaining U.S. economic competitiveness" to concern over the perception of corporate welfare to questions regarding the industrial globalization and the possibility of U.S. taxpayer dollars supporting foreign entities. Less debated, but equally important, has been the national laboratories' potential competition with academia for federal research dollars and discussions detailing the role of each in the national research enterprise.

Industrial Collaborations

The Department of Energy's (DOE's) Technology Transfer Initiative (TTI), which grew rapidly from 1991 through 1996, was curtailed by Congress in FY 1997. Although few people would wish to replicate the TTI program, at the time it was particularly beneficial to the laboratories because it created a stimulus for cultural change, both in industry and the laboratories. Relationships were forged, visions were shared, and each side learned about the other. For the laboratories, the initial industrial collaborations were a learning experience and an awakening to the outside world. The laboratories found that industry's respect was not easily earned. The laboratories had some early failures: They had not fully appreciated the difficulty of transforming research into commercializable material; they did not always listen well or understand their partners' needs; and they met many other smart people who knew more about some things than they did. Conversely, industry found out that the laboratories were neither totally removed from the real world nor ignorant of the important technical issues confronting industry. Industry learned what the laboratories and their primary customer, the DOE, valued and what pressures the labs faced in a time of shrinking federal research budgets. Industrial partners found some very exceptional scientists doing work that was surprisingly relevant to industry.

With the discontinuation of TTI and the easy money that flowed with it, the partnerships created between industry and national laboratories needed to have more value to each partner in order to justify the full cost. This is the time period when fruitful industrial–national laboratory collaborations really began to flourish. Congress ended TTI because it wanted the federal share of joint work to be funded out of mission programs rather than as set-asides. The post-TTI model ensures that the work is of value to a federal program manager and is not merely "welfare" to the laboratory or the corporation.

Like many of the other national laboratories, Sandia National Laboratories is committed to maintaining its scientific excellence through continued collaborations with industry and academia. The most shining example of a Sandia and industry collaboration is the longstanding cooperative research and development agreement (CRADA) that Sandia has had with the Goodyear Tire & Rubber Company. Together, Sandia and Goodyear are developing and validating tools for finite-element analysis for predicting thermal and mechanical responses of structures. The CRADA also includes using a variety of analytical techniques to study the structure and properties of various materials used in tire fabrication to gain an understanding of the aging and reliability of elastomers and developing methods for predicting material lifetimes. For Goodyear, these tools can be used to simulate and predict manufacturing elements such as shaping and curing processes and performance characteristics such as rolling tire resistance and hydroplaning. For Sandia and DOE, the research performed for predicting materials lifetimes is highly synergistic with work being done to support Sandia's science-based stockpile stewardship mission.

Nissim Calderon, Goodyear vice president of corporate research, as quoted in the October 1996 issue of *Mechanical Engineering*, explained the dual benefits in algebraic terms. "You greatly enhance the chances of success by making it a win–win situation for both the lab and the company. Say Goodyear has a two-component project, A and B, while Sandia has another project characterized by A and C. Why not do A together and share the risk, share the effort, share the cost? And while we're at it, their scientists rub elbows with ours, and naturally they exchange expertise and suggest changes in current procedures, so the laboratory is also acting as a consultant, while the lab personnel get a reality check from business."

With this as its prototype for laboratory–industry research, Sandia actively encourages industrial interactions that are mission related and enriching to the scientific base of the missions and to Sandia's technical staff. Sandia, in both its activities and structure, supports partnerships in a variety of ways. For example, Sandia is the major contributor to the development of an industrial research park located directly outside its gates: the Sandia Science and Technology Park. In another program, individual Sandia executives are assigned "ownership" or responsibility for each major corporation with whom Sandia does business, giving the company a high-level contact at Sandia who can smooth and hasten communication by acting as ombudsman, advocate, and negotiator.

University Collaborations

Regarding its relationship with academia, Sandia recognizes that partnering with universities is essential for maintaining Sandia's excellence in state-of-the-art science, for being able to tackle and solve complex problems, and to maintain the ability to recruit the best new scientists and engineers. Partnering with universities is an official strategic policy—Sandia has chosen to focus on a few key campuses (about 20) and has made a significant effort to build a relationship between the leaders of the universities and the Sandia executive management team. Top management at Sandia are campus executives, and each is tasked with developing a high-level relationship with a specific university. The universities that Sandia chose to focus on included those in which Sandia had a strong past experience of research interactions, a successful history of recruiting, effective institutional ties, and common future interests such as computing, research, diverse work force opportunities, and the use of information technology in teaching.

To further university interactions, particularly with Sandia's closest neighbor, the University of New Mexico (UNM), Sandia supports the Advanced Materials Laboratory (AML), a UNM building situated next to the campus and housing about 25 Sandia staff, about a third of whom have joint appointments at UNM. The building is wired for both the Sandia and the UNM networks, is accessible to foreign national students, and is the center for materials collaborative research with UNM. Although the AML houses about 1 percent of Sandia's Ph.D. staff, about 11 percent of Sandia's refereed publications are from AML staff. The AML is filled with students and visiting faculty from both the United States and abroad.

Catalysis

Catalysis and the Sandia Mission

Would a national security laboratory whose primary mission is stockpile stewardship be interested in supporting catalysis research? Catalysis interacts with and supports Sandia's missions in a number of critical ways. Catalysis and chemical reactors benefit from advances in three of Sandia's primary research foundations that form the laboratory's core competencies: materials and processes, engineering sciences, and computational and information sciences. Breakthroughs in ceramics, surface science, hydrodynamics, high-temperature chemical reactions, new porous materials, computational techniques (micro-, meso-, and macroscale), engineering diagnostics, and many other research areas that Sandia invests in for its nuclear stockpile stewardship mission can be used to impact catalysis in profound ways. In addition to the stockpile stewardship mission, Sandia also has a mission focused on advancing the surety (safety, security, and reliability) of critical infrastructures, including energy. Clearly, catalysis contributes to energy security in a number of ways: Most chemical production is dependent on petroleum-based feedstocks that are synthesized using catalysts; catalysts allow the production of liquid transportation fuels from alternative sources such as natural gas or biomass; and better catalysts produce more efficient industrial production of chemicals leading to less environmental waste, greater economic viability, and less energy usage. Catalysis is also a tool for forming partnerships with the chemical and petroleum industry. It is this intertwining of catalysis and chemical reactor design with the core competencies and energy security mission of Sandia that makes it a technology that meshes well with Sandia's mission.

What motivates a program, such as the catalysis program at Sandia, to actively pursue partnerships with industry and academia? In general, the answers are not too different from what motivates Sandia National Laboratories, as a whole, to partner: to leverage federal funding (if we need A and B and you need A and C, let's do A together) and to maintain technical excellence. Rubbing elbows with industrial scientists and engineers helps keep Sandia researchers cognizant of industry's needs, information that is crucial to the study of catalysis.

Teaming and Partnering in Catalysis Research

During times of shrinking nondefense federal research dollars, the motivations for partnering become more complex, driven at times by a desire simply to survive. In the field of catalysis, this drive for survival is especially keen. Federal spending on catalysis research has decreased consistently for many years. International government investment in catalysis research has made the global market for industrial catalysis research funding highly competitive. DOE national laboratories compete directly with academia for federal funding dollars in catalysis. When in the single-investigator mode, laboratory researchers are likely to fare poorly against academia, because laboratory overhead is much greater than most university overhead. The ability to team internally, between laboratories, with industry and with universities, particularly in a multidisciplinary fashion, is a factor that can add significant value to laboratory-based research.

Recently there has been an effort for the labs to work more closely together in catalysis.[1] This effort is motivated by a desire to raise the level of awareness about catalysis within the government and to educate industry more about the capabilities within the national laboratories. One of the critical motivators for collaborating with industry is the need to develop industrial advocates with Congress to maintain support for catalysis research and capabilities in the United States. Several European countries have recently invested a significant amount of funding into their catalysis research efforts. The British have recently formed the Institute of Applied Catalysis, a virtual institute of catalysis that brings together industry and academia and supports (financially) research. Through the Netherlands Institute of Catalysis (NIOK), catalysis has been one of a few disciplines to get major support from the Dutch government. It too is a virtual institute that works closely with industry and benefits from significant government financial support. Professors from seven universities in the Netherlands participate in one graduate program in catalysis, and students can take classes from any of the universities. The German Institute of Applied Chemistry Berlin-Adlershof (ACA), where catalysis research is performed and supported by the German federal government, state of Berlin, and industry, has grown by 20 percent in recent years. The French government also supports catalysis research at a singly devoted institute: France Institute of Catalysis Research in Villeurbanne, France. It is against these well-supported institutes with well-maintained infrastructures that DOE national laboratories and U.S. university catalysis researchers compete for industrial support. U.S.-based chemical companies are frequently found as active and financially supportive members of the European catalysis institutes.

To be more attractive as collaborative partners to U.S. industry, the national laboratories are trying to make their capabilities better known to industry. At Sandia, an effort to tie in diverse parts of the laboratory to help potential partners solve their problems is a critical aspect of our strategy. For example, when investigating a catalyst intended for use in a slurry bubble column reactor (SBCR), we learned that our customer was also interested in learning more about the hydrodynamics of an SBCR. Over the course of several years, Sandia built an industrial-scale, nonreacting SBCR and developed diagnostics for conducting research into three-phase hydrodynamics. The SBCR was built where it could have access to the air storage tanks for the wind tunnel facility, which allowed Sandia to test high-velocity flow rates in industrially relevant diameter reactors. This is an experimental setup that would never be possible in a facility that had to depend on gas cylinders as a gas source. Sandia has also tapped into their surface and materials scientists' skills to expand new surface techniques to the study of catalysis and develop new materials and membranes to catalysis and membrane reactors. However, obtaining funding for carrying out much of the more fundamental aspects of catalysis research, despite Sandia's strong support of the catalysis effort, has been quite challenging.

Funding for Catalysis

Funding sources for catalysis research in the United States is quite dispersed and may be found in several divisions of the National Science Foundation (NSF) and four or five groups within DOE, the

---

[1] Mitch Jacoby, "Planning for the Future at DOE Laboratories," *Chemical & Engineering News*, March 29, 1999, p. 25.

largest being the Chemical Sciences Division in the Office of Science; and the fastest growing is the Industries of the Future Program in the Office of Industrial Technology. The Environmental Protection Agency supports a small amount of research in its Green Chemistry efforts. The American Chemical Society Petroleum Research Fund, a private funding source, also supports a small amount of catalysis research at universities. The scattered approach to catalysis funding leaves catalysis without a strong advocate within the federal funding structure. This is in strong contrast to the situation in the United Kingdom and the Netherlands, where the Institute of Applied Catalysis and NIOK, respectively, are strong and successful advocates for government funding of catalysis.

Of course, industry also funds catalysis research both internally and externally. However, the closing of Mobil's highly acclaimed corporate research laboratory in Princeton, where zeolites were first developed, as well as closings at many other corporate laboratories, the restructuring of the petroleum industry, and the increasing focus on short-term research have led to a significant decrease in catalysis funding within the U.S. industry.

Teaming and Partnering—Its Barriers and Drivers

Because of the proprietary nature of industrial catalysis, there are many barriers to industry "outsourcing" catalysis research, particularly the development of a specific catalyst. A good catalyst provides an economic advantage to its company only when the catalyst formulation and preparation remain secret. However, industry has identified many fundamental research issues in catalysis that, if they were to be addressed, would have a significant impact on the economic vitality of the chemical industry.[2] Catalysis research and technology needs were the subject of a workshop held as part of the Vision 2020 technology roadmapping process. Vision 2020: The U.S. Chemical Industry is a project sponsored by five chemical associations to identify areas requiring research in order to maintain a healthy and vital U.S. chemical industry into the year 2020 and to "roadmap" a process for accomplishing the research.[3] However, the chemical industry has not entered into government–industry–university collaboration to address these fundamental research needs as, for example, the electronics industry has in precompetitive projects like SEMATECH. The reason for this lack of collaboration may stem from a historically more adversarial relationship between the government and the chemical industry than the government and the electronics industry. In Europe, where a number of countries have institutes (real and virtual) devoted to catalysis, the relationship between government and industry, including the chemical industry, appears to be different than in the United States. In Europe there is no concern about "corporate welfare," and the relationship between government and industry has been likened to the relationship in the United States between the U.S. government and the defense industry. Vision 2020 is an effort to improve the relationship between government and the chemical industry. Industry shares information regarding the technology needs it perceives as necessary to ensure a vital U.S. chemical industry in the future. Government can, and has, supported research in these areas, more effectively and efficiently spending tax dollars in support of a future robust economy.

In summary, for a national laboratory contemplating maintaining a rich and flourishing catalysis research effort, the challenges are significant. Overall federal funding for catalysis within the United

---

[2] N.B. Jackson, *Catalysis Technology Roadmap Report*, SAND97-1424 UC-1404 (Albuquerque, N.Mex.: Sandia National Laboratories, 1997).

[3] American Chemical Society, American Institute of Chemical Engineers, Chemical Manufacturers Association, Council for Chemical Research, and Synthetic Organic Chemical Manufacturers Association *(Technology Vision 2020: Report of the U.S. Chemical Industry* (Washington, D.C.: American Chemical Society, 1996).

States is decreasing along with many other types of nondefense, energy-related federal research funding. A national laboratory must technically distinguish itself from its academic colleagues, because the cost of national laboratory research is higher than university research. Because of the strong government support of catalysis research in Europe, a U.S. national laboratory has significant competition globally for industrial investment in catalysis research. There is no strong advocate for federal funding in catalysis research because there is no one primary funding agency, and U.S. industry, historically, has not been an advocate in this area.

There is a remarkable amount of excellent catalysis research being conducted in the United States. Unfortunately, at many of the national labs, there is a lack of "critical mass" in catalysis. Many have perhaps only one to three full-time catalysis researchers along with a few surface scientists, microscopists, and/or synthetic chemists who may include some projects related to catalysis in their portfolio. The effort to get the national laboratories to work closely together is an effort to create a virtual catalysis community where "critical mass" is achieved. It is also an effort to leverage and capitalize on programs technically related to catalysis, to fully utilize all DOE materials facilities (such as the synchrotron at Brookhaven and neutron diffraction at Los Alamos), and to provide a coherent program that will attract industry's interest. As a team, the labs can also be more effective in raising the profile of catalysis research. If industry finds the laboratories' catalysis research useful, it will also likely become an advocate of catalysis research support.

Universities have also created catalysis centers to better address industry's needs and to provide a more coherent program for students. One of the most innovative of these programs is Northwestern University's NSF-funded environmental catalysis program. In this promising program, industry is invited to participate actively in the education of graduate students and the research they conduct. The national laboratories are partnering with Northwestern University and other university catalysis programs. This teaming makes for a strong package of catalysis research and benefits to industry, including well-prepared graduates.

CONCLUSIONS

What slows down this process of teaming in catalysis among laboratories? Certainly a culture of scarcity and uncertainty is the most significant contributor to an aversion to teaming or sharing and a desire, rational or not, to hold on strongly to what little has already been granted. Conversely, an entitlement attitude (undoubtedly closely related to what is sometimes called a "culture of arrogance") also discourages teaming and partnering. A researcher whose laboratory already has sufficient funding and perceives a guaranteed future support for catalysis can easily continue to work in a single-principal-investigator mode. Why would that principal investigator want the difficulties or lack of control associated with a research team or a virtual laboratory?

What discourages industry from working with the laboratories or being an advocate for catalysis research within the laboratories? The difference in culture between industry and government is a significant problem. The bureaucratic burden put on a company contemplating a partnership or CRADA with a laboratory has decreased since the beginning of the 1990s but is still far greater than industry (or government) would like. Political uncertainty is always an issue. For example, industry–laboratory programs have come and then quickly gone, like the TTI. In general, the lack of a clear mission for the laboratories (beyond stockpile stewardship) contributes to the uncertainty regarding future funding and direction for the labs, increasing the risk of a long-term industrial collaboration.

Although there are many barriers to teaming among the labs and working with industry in the field of catalysis at present, particularly with the competition from Europe, the continued and sustained excellence of catalysis research at a large number of labs will be obtained only by working together. The labs, along with university partners, need to differentiate themselves from their competitors by providing a virtual catalysis research community that will offer multidisciplinary research, both fundamental and applied. Only by appealing to industry will the laboratories be able successfully to increase the profile and advocacy for catalysis within the United States.

ACKNOWLEDGMENTS

Many thanks to Dan Hartley, vice president, Sandia National Laboratories, for his advice, his willingness to share his insights, and his relentless support of partnerships.[4] This work was supported by the DOE under contract DE-AC04-94AL850000. Sandia is a multiprogram laboratory operated by Sandia Corporation, a Lockheed Martin Company, for the U.S. Department of Energy.

DISCUSSION

Donald Burland, National Science Foundation: I am familiar with the Dutch situation. One of the things they do that we don't do is to realize that, as a small country, they cannot do everything, so they pick key areas. Maybe we have this problem that we do everything, and we don't make choices. When scarcity comes along, we still try to do everything. I wonder if there isn't some incentive to look at trying to do less and to do it well.

Nancy Jackson: I agree 100 percent. The answer for the national labs in the nondefense areas is to somehow choose the specialties and the expertise that each laboratory is going to have, which does not mean that only one lab is going to be catalysis. It just means that one laboratory might be better at surfaces, or one laboratory might do materials, or there needs to be a closing down of the areas that we get involved with.

In the beginning of the TTI program, there was a feeling at Sandia that, as scarcity began within the labs, there was an "any dead cat" mentality. That is what we called it. You drag in any dead cat for funding because it was a matter of staying alive.

We have tried to stay away from that because it is a knee-jerk reaction, not a long-term strategy. And each laboratory needs to decide what is important, or there needs to be some upper-level decision making.

In the Netherlands I was amazed to see that they ranked every professor in the system from one to five. With five being at the top, they were eligible for certain types of funding and nobody else was. This will never happen in the United States. So they are capable of doing things that we aren't. I hope it will never happen in the United States.

Randolph Guschl, Dupont: One of the fears that some of our management have is that, in catalysis, someone will do some research and discover a new catalyst that will take away a whole business. Have there been any documented cases in which that charge has been leveled at your programs?

Nancy Jackson: People say that to us all the time, but we don't have an example of it. Most of the funding agencies within the government don't fund us to develop a catalyst. They fund us to understand

---

[4] D.L. Hartley, *The Future of the National Laboratories*, SAND97-2556C (Albuquerque, N.Mex.: Sandia National Laboratories, 1997).

why the carbon deposits on the catalyst or this type of thing. And so, yes, you might come up with a good catalyst in the meantime, but we would have to team with a company in order to make it real.

Andrew Kaldor, Exxon: This discussion on catalysis is a good example of the gulf in the scientific community in the United States. We will be facing fuel regulations in 2004, which will require low-sulfur fuels. The debate on this has been going on for ages. We can do it today, but it will cost a lot of money to do it.

It would have been such an ideal opportunity for the Environmental Protection Agency to work on the regulatory part on one side of the government and the DOE to work the problem from the other side. The National Petroleum Institute says that you cannot do any of this. It would be ideal to say, "Catalysis is going to cost $100 billion to create this regulation in terms of investment in the United States. We'll develop the technology to cut that in half." It would be such a wonderful opportunity.

As you look ahead, is this really not a basis for going forward? We have found that virtually no technology exists at the universities in the United States or national labs that actually solves the problem on the time scale that we have to solve it. We have to go outside the country. Even with that there are limitations.

I am just raising this as a paradigm on these partnerships. Shouldn't the regulatory agencies, which are setting the targets, work with the other arms of the government to either encourage private industry early to find technical solutions or participate in them?

Nancy Jackson: Nowadays, industry has to say it first, and then government will do something about it. And then they'll give the labs and the universities the money to do it. But government is not, particularly in any energy-related area, going to take the ball first. No one cares. It's not important enough to the government. Gas is cheap, chemicals are bad. I could go on and on. Unless industry says it first, it's not going to happen.

James Desveaux, University of California, Los Angeles: You sound as if you have resigned yourself to a retrenchment strategy for the labs, which is a bit worrisome. You have just drawn some comparisons between Europe and the United States. One strategy is to draw in our horns and focus more on those things that we do really well. Regarding labs, this amounts to a strategic focus on comparative advantage. But from the political side, my fear is that this might result in a downward spiral for the labs, because one never knows from one Congress to the next what catches their eye and what incenses them. It seems to me that an alternative strategy might be to make the case—and I realize it is a difficult and challenging one to make—that the national labs are national resources and ought to be protected and supported as such. Why not carry the debate at a higher level of analysis, which might also protect you from some of the accusations that these labs are exploited, for instance, for corporate welfare.

Nancy Jackson: I'll go back to what I said before. Industry has to say it first. My management cannot go to Capitol Hill and say, "you know, we're great. We're a national treasure." It has to come from outside, and I don't see that happening.

It goes down to the basic relationship—I am only cognizant of the chemical and petroleum areas because of my background. But because of the historic adversary relationship in that area, we don't have industry as advocates for the labs. That is why I have been so interested in Vision 2020; we need to have industry as advocates.

There's a risk no matter what we do. Maybe I spoke negatively about this retrenching. How I really see this working together is as a joint effort on catalysis. I see it as an opportunity for the entire field of catalysis to grow in an interdisciplinary way. We could do this as engineers, chemists, and physicists with our incredible equipment. I see it as a rebirth of my discipline and a very exciting opportunity in the future.

John Tao, Air Products and Chemicals, Inc.: Politics aside, what are the process and the time frame to define and focus on a mission for the labs?

Nancy Jackson: We have been doing it for ten years now, and I don't think we have gotten much closer than we did before. At my own laboratory my management struggles with this. But they take it on and they make a decision as to what Sandia is going to be good at and what they're going to do. Ultimately they could be perfect. They could do a fabulous job, but ultimately it is Congress and the government that make the decisions.

Sandia and perhaps many of the other labs have done a good job in defining their missions. Internally we understand what that is. But whether this is recognized from the outside, I don't know. It's beyond my control. To a certain extent, it's beyond my management's control too.

Thomas Manuel, Council for Chemical Research, Inc.: First, the oversight that Andrew Kaldor discussed might have been remedied had the petrochemical industry become involved in the industries of the future roadmap exercise when it was first proposed, which was more or less coincident with that timing. As you said, the purpose of the roadmaps is to identify futuristic targets and provide some ammunition and support for research in those directions. So there is a mechanism.

Second, as far as gaining influence with Congress, along with calling your local representative and seeing him or her in the local office, there is a Congressional Visits Day, which is organized every year by coalitions for science and technology partnerships and for science funding, that musters over 200 individuals representing organizations that we all belong to under an umbrella group to descend on the Hill and advocate science funding and other subjects related to science that interest us. So there are mechanisms in place to influence the congressional position, and it is up to the industrial and academic people to participate in them. The Council for Chemical Research is one of a hundred organizations that participates in that effort, but we, other than the federal employees, need to get involved. It's not creating something from nothing. There are processes that are struggling to get momentum and that exist to provide this support.

Hank Whalen, PQ Corporation: You are right. Somehow or other industry has to get with it and do something, because otherwise the gloom and doom that you paint could happen.

Bruce Harrer, Pacific Northwest National Laboratory: I have heard rumors that the TTI program is completely going away, and I am concerned that partnerships will fall dramatically. Henry Kohlbrand of Dow Chemical talked about the desirability of having someone coming to the lab, finding out what they're doing, and finding a program linkage. If there's no flexible mechanism for the labs to get funded to go forward with these partnerships, what do we do?

Nancy Jackson: I agree. I just hope that it can be understood that it's essential to our mission, even our national defense mission, to have interaction with industry. That is the only way of convincing people that if we are going to have the best weapons or security systems or whatever, we have to be connected with the industrial community to know what the latest and greatest is and what to add or what to not add to our weapons. That is the way I see our best argument.

Todd La Porte, University of California, Berkeley: Let me make an observation followed by an irrelevant question. I'm struck by what Nancy Jackson has shared with us. There is a sense of vulnerability of the very good analytical institutions in the United States, that we have come to expect to be remarkably good whatever happens to them. And no matter how good these institutions are, there is a willingness on the part of industry to rationalize its disregard for our society in terms of globalism. This is really extraordinarily apparent in this room today. That willingness to say, "we can go outside (overseas) and get it cheap and become international, that is, to withdraw political support from our own analytical institutions, not just Sandia."

But we know that the very same thing, not in specifics, but in the general pattern, is occurring at Los Alamos, at Livermore, and at Lawrence Berkeley National Laboratory as well. There is a question here with regard to American industries' understanding of its role in our national security interests in the long term.

Politically, corporations have moved in the direction through Congress of cutting the capacities of our public institutions with regard to analysis. I understand some of the background for it and some of the ideological drivers. But what Nancy Jackson has told us is an example of this decline in the willingness in our political culture, and I mean its industrial cultural element, to support national analytical capabilities that are not immediately responsive to particular individual corporate interests. We could lose the whole thing over time unless there's a much greater sense of corporate stewardship as well as national stewardship with regard to these matters (supporting national analytical capacity). And I do not hear this conversation taking place anywhere. It's a very serious problem.

And now my question. How does Lockheed Martin use its management of Sandia to help its own corporate intelligence with regard to technology? Does that work, and if so, how?

Nancy Jackson: I am not comfortable commenting on that, except that, as you can imagine, they are more closely in touch with the kind of science and engineering that we are doing at Sandia because they physically work with it. And so they benefit just intellectually, if nothing else, from that type of interaction.

Todd La Porte: You are doing better than the University of California.

Panel Discussion

John Ekerdt, University of Texas: My question is for Andrew Kaldor, who is a representative of the petrochemicals sector. I want to follow up on the presentation by Todd La Porte describing how the Plant and Microbial Biology Department at UC Berkeley entered into a partnership because it needed access to the agricultural genomic database that Novartis had. Let's take this idea of desulfurization and say that the federal government mobilized laboratories and funding to people working on this problem. Will the industry, knowing what they know after over 50 years of working on desulfurization, open their vaults so that for the first ten years of the partnership people don't reinvent what has already been learned? What is the likelihood of this happening?

Andrew Kaldor, Exxon: At this point I can't give you an answer. Four or five years ago, the general position for much of the petrochemical and chemical industry was that environmental-related technologies were going to be used by everyone and that the competitive edge for environmental technologies was not very practical. So this was one example of a rather insular industry cooperating to produce an unfair, unlevel playing field solution. There was no way that the smaller companies, some of them oil refineries, could make the investment required to meet the regulations. As a result, the bigger companies, like Exxon, are forced to toe the line, and the smaller companies get stretched out over several years. Had the technology been broadly available to everyone at reduced investments, everyone would participate at the same time, and the excuse of working for a small company would no longer hold up. So the discussion could have taken place.

There was a situation in which the Environmental Protection Agency was heading toward regulations for very low sulfur levels, which was strongly supported by the automobile and engine manufacturers. Industry kept complaining about the huge investment that was required. One of the big companies said that they agreed with the investment requirement, but noted that if the problem could be solved by another route, that investment could be significantly reduced. This could almost start a national program on the basis that the objective is to reduce the cost of meeting the regulatory environment. I believe there is still an opportunity to worry about this because I don't believe that the current regulations are the final ones. We will probably head toward having no sulfur in the longer term. I am discussing this more from the point of view of a paradigm shift. Had such a discussion occurred, it probably would have made the companies open the coffers. Exxon has spent close to $40 million to fill the coffers because they were so bare.

Ashok Dhingra, Dupont: If you look at the past ten years of market economy, globalization has driven some of the investors such as textile, steel, and smoke stack industries overseas. The result has been that the United States has concentrated more on a highly technological knowledge economy and has worked well in general with a fcw bumps here and there. If this happens in the area of research—if catalysis is better done in Holland, England, or Japan, and similar technologies make sense—then what is wrong if the multicultural and multinational companies move overseas?

Nancy Jackson, Sandia National Laboratories: Coming from a security-oriented national laboratory, the first thought that I have is that the United States does not want to remain vulnerable regarding some technologies. It does not want to be dependent on any more foreign sources than it already is. We are already very dependent on foreign sources for oil. We certainly don't want to lose any key technologies as well. This could be construed as a national security issue. Perhaps catalysis is not important enough to be part of that, but certainly some information and other technologies will be.

Ashok Dhingra: That sounds like a self-serving or survival game. Take carbon fiber, for example, upon which the Department of Defense is heavily dependent for security systems. Look at the big supplies in this country, and you can see that it was economically more sound to get carbon fiber from Japan. So this debate has gone on for a long time. Market forces gave the sign that Japan was the best supplier of carbon fiber.

Nancy Jackson: This is a political decision. What the government wants from internal sources versus external sources is a political question, not a technical or researchers' decision.

Ashok Dhingra: It's a market decision.

Nancy Jackson: The options are driven by the market, but which materials or technologies we are to rely on is the decision that needs to be made politically.

Henry Kohlbrand, Dow Chemical Company: I tend to agree with Nancy Jackson. There are two elements that we have to look at: the industrial component and the political context. They are not easy to mix. It is easy to say that we should look at the nationalistic responsibilities of American companies in assessing how we should deal with this. However, defining what an American company is, in today's world of multinational companies, is difficult. One could define an American company to be one that was founded in the United States and then expanded to other parts of the globe. Today, in the chemical industry, the largest companies are European companies such as Bayer, Badische Aniline Soda Fabrik Corporation, and Hoechst. Even though they have significant operations in the United States, they did not originate here. What is their responsibility to support the national interests of the United States? These industrial and political issues need to be addressed separately and then an attempt should be made to bring them together. This will not resolve itself.

Joseph Cecchi, University of New Mexico: Maybe we should explore how general this situation is. I am familiar with Sandia partnerships in microelectronics, and they have been quite strong. That is not to say that every one has been successful, but realistically you wouldn't expect to measure success by having every interaction succeed, but rather by the long-term viability of interactions. Nancy Jackson mentioned the Extreme Ultraviolet Lithography Consortium, which is certainly the most prominent. And I think industry is going to Congress and saying, "We need the laboratories for this." As a side bar on that, it is interesting to look at the characteristics of that program because it has generated a lot of controversy because of the scale, the exclusivity—basically just a few companies and three labs—and the duration. So there are probably aspects of that this could be seen to overlap with the Novartis case. Could you comment more broadly on these other areas in which maybe Sandia has been successful, or perhaps it is just that certain areas have a more natural overlap with industrial interest? If you look more broadly, it may not be as dire from the standpoint of the federal laboratories.

Nancy Jackson: I agree. I tried to allude to that in the sense that historically the chemical and petroleum industries have had more of an adversarial relationship with the government. And that is not true as much with the microelectronics industry. As a consequence, there is a very strong interaction between the national labs and the microelectronics industry. Certainly they have a long history of that. And your point is good, that in those areas in which industries have a less adversarial relationship with the government, there is a very good, strong, and healthy interaction with the labs, and with the universities too, for that matter.

Joseph Cecchi: Just to follow up, you make quite a point of the factor being the adversarial relationship with government. Are you sure about that? Could it not be that there is just somehow more of an overlap in which industry finds that they are getting something in areas of microelectronics at a precompetitive level? I'm wondering if it has to do solely with the relationship of industry to government, which would be unfortunate, I think, or whether there are aspects to it that just have to do with the details of that particular need of that industry and what they find at the national level.

Nancy Jackson: I think that it is both. Within the chemical industry, certainly in the areas in which there are more precompetitive issues, we have seen in the past five or six years, certainly with Dow, that they have worked closely with the national labs in these precompetitive issues. The computational fluid dynamics consortium is a very good example. Separations is another good example; often there is a lot of precompetitive work in that area. There are large consortia in that area, not with Sandia, but with other national labs. But Sandia is doing a lot of work with industry on separations too. Yes, there are areas other then catalysis that are farther along with the chemical industry, but I am unsure if they are as far along as microelectronics.

Frank Feher, University of California, Irvine: I'm not doing catalysis research now, but I was trained to do catalysis research and I have actively followed it for many years. I am very uncomfortable with your characterization of catalysis research and the implications it might have outside of Sandia. The United States does not have the big centers for catalysis research that the Dutch have, but if you look at all of the major developments in catalysis, you will see that a good case can be made that the United States is still very much a leader in the field. And that is not likely to change any time soon. So Sandia may be looking for partners and trying to build up centers to survive, but it's unfair both to the other researchers in catalysis in the country and to the researchers at Sandia to paint such a bleak future for the present and future of catalysis.

David Schetter, University of California, Irvine: If Dow were to be given agreed-to royalty ranges up front for exclusive license and under certain conditions nonexclusive royalty-free licenses for other work that was more incremental or less a product in and of itself, how much of the intellectual property concern would that solve for Dow?

Henry Kohlbrand: This would need to be evaluated on a case-by-case basis. We do have some blanket agreements (and have established common principles for these agreements). But if you are talking about a project much larger than the support for one graduate student, a number of questions would be raised. What is the expected outcome? How strategic is the project? Is it precompetitive like some of the examples that Nancy Jackson gave? Depending on the answers to these questions, you can come up with a number of cases in which simple formulas will work very well. However, if we are talking about a development that is in the middle of a very important strategic technology development for the company, there would be less flexibility. If you look at the way we are capitalizing on recent techno-logical developments, it is not by simply starting businesses within the company but through joint ventures or new companies involving several partners. Will a royalty-free exclusive license be able to be used in these more creative business constructs (less than 50 percent ownership by the primary company)? Can it be transferred easily to the new business entity? So what appears simple can be very complex. Our flexibility in negotiating terms for intellectual property is more a function of where the technology fits into the overall business strategy. If it is at a development stage, we tend to be very flexible. If this is close to or in commercialization, we are much less flexible.

Michael Chartock, Lawrence Berkeley National Laboratory: I would like to reiterate some of the strengths of the nation's scientific research in the area of catalysis. In our laboratory, the Department of Energy (DOE) has made major investments in terms of the Surface Science and Catalysis Laboratory; upgrades to the National Center for Electron Microscopy; and constructing, building, and operating the Advanced Light Source, which is very much dedicated to research in the science of catalysis. And so I see in many ways a real investment in this area of science. With the neutron source, which can do both bulk materials and look at surface properties as well, we have a tremendous opportunity. Perhaps the decreased operating budgets of the DOE and the limited increase in funding at the National Science Foundation don't compare with the redirection of the nation's scientific expertise in life sciences that the national institutes have garnered. But overall there is a tremendous positive view and even support for the laboratories' mission. In our laboratory we do have a very important, strong mission, and the Office of Science at DOE strongly supports that.

Nancy Jackson: My presentation must have come off more negatively that I meant. I actually agree with you. I believe that the United States has good deal of exceptional research. What I am trying to say is that we have an opportunity, if we partner and do multidisciplinary research, to be truly extraordinary. When I talk to some of the people in Europe about a potential virtual laboratory or a collaboration within the United States, they know that we would be formidable competitors if we were to organize in the same way that they are.

Henry Kohlbrand: I would like to make a comment on the subject of doing research in the United States versus elsewhere. In the Dutch example, is appears that they have been very sensitive to the competitive situation in catalysis research globally. They have defined gaps that they are trying to fill and they are very open to collaboration. Sometimes in the United States we appear to be racing against ourselves. Rather than following the Dutch example and defining areas of work for each institution, we have a tendency to have several institutions running very similar programs. This has created some redundancy in capability and duplication of effort and has consumed resources that could have been used to help one institution reach a high level of capability. In the globally competitive environment in which we operate today, we can afford to do less of this than we have in the past. In the catalysis example, we should consider better organizing the efforts across institutions. We need to be sure that each institution understands what the other organizations are doing, what their strategies are, and how they fit into the overall picture. Hopefully, we can then assemble an effort that fits together smoothly rather than pulling against itself.

Christopher Hill, George Mason University: Earlier there was a dialogue on the causes of the relative collaboration levels between the chemical industry and other industries with the federal laboratories. In the ultraviolet lithography case, Berkeley's light source and Sandia's semiconductor manufacturing facility are extraordinary and unique capabilities of the laboratories that nobody else has and that make them critical partners in that cooperative research and development agreement (CRADA). What was not mentioned, of course, was the intention of one of the partners to transfer the resulting technology to Japan. This is where the controversy came from, not from the project's scope, scale, budget, or commitment of time. It is also true that the chemical industry was quite reluctant to join in collaboration. And, relative to other sectors, the chemical sector is new at the table. I was a member of the American Chemical Society Committee on Chemistry and Public Affairs where the first dialogues took place that led to Vision 2020. With the exception of one or two leaders from industry, the industry people were very reluctant. The Chemical Manufacturing Association's lawyers were very hesitant. That is something that other industries got past a long time ago. And so now we have some catching up to do.

Lura Powell, National Institute of Standards and Technology: That is an excellent lead-in to my comment. The Advanced Technology Program (ATP) covers a breadth of technology, and as director I see a similar thing occurring in our program. The chemical industry has been more reluctant to partner with the ATP than other industries. In fact, there wasn't much chemical industry participation at all until the catalysis and biocatalysis program started in 1995.

I would also like to raise an issue about roadmapping for the chemical industry. I have looked at many other industrial roadmaps, including those done in electronics and optoelectronics, in which they have been able to identify the science and technology advances needed to move the entire sector forward. These industries have been able to use their roadmaps very effectively to foster industry collaboration and leverage federal funding toward their goals. In the Vision 2020 document, the global vision is really good, but it needs to include more details about the science and technology research needed to realize the vision. Vision 2020 does not yet spell out what is needed to create the chemical industry of the next century—certainly, not in the same way and with the type of organization that I have seen in other roadmaps.

One of the issues that the Chemical Sciences Roundtable could grapple with, and that would advance this effort, is to look at the profile of collaboration and partnering in the chemical industry versus other industries. What can be done to take better advantage of the partnering opportunities that exist across industries and that include scientists from universities, national labs, and not-for-profits? I see collaboration as a tool for achieving real advances in the future. I think we have heard that from others as well.

Henry Kohlbrand: I would like to comment on some of Nancy Jackson's observations. One of the frustrations that I have had with Vision 2020 is the scope of the definition of the chemical industry versus the electronics, aluminum, or power industries. If we ask what constitutes the chemical industry, we come up with a number of different answers. To one person, it might be a group of specialty manufacturers with relatively small, geographically distributed facilities that make small-volume, high-value products such as pharmaceuticals. To another it might be petroleum refining and commodity chemical production in very large facilities, to another it might involve plastics, to another fertilizer manufacture, and so on. The way that we define "chemical industry" is very broad and covers thousands of processes and a large number of process technologies. It is probably unfair to expect that we will have the same focus for Vision 2020 in the chemical industry as we do in the aluminum or micro-electronics industry. In aluminum and microelectronics, there are a small number of processes and fewer unit operations than we have in the chemical industry. Perhaps we need to divide the chemical industry into logical subgroups in order to address our concerns in a more orderly way. We have focused on some areas of technology in the chemical industry with extraordinary depth and virtually ignored others in our roadmap activities. Although we have made a lot of progress, there is more yet to be done. I don't think we have been fair in our expectations of the chemical industry, and we may need to take a different approach than we have in the more focused industries that have fewer processes.

Cheryl Fragiadakis, Lawrence Berkeley National Laboratory: Would William Millman be willing to make any comments about the importance of partnerships now and in the future for the research that he is directing? The question is directed more to the Office of Science because it is such a major factor in the research that we are talking about here. How do you measure their progress?

William Millman, U.S. Department of Energy: I cannot speak in general for the Office of Science, but I can certainly speak for the Chemical Sciences Division and maybe more generally for Basic Energy Sciences. As Nancy Jackson mentioned, the research budgets for the department over the past ten years have not kept up with inflation. As a result, funding at the laboratories has not kept up with inflation. Therefore, we have encouraged the programs that we support to look for other sources of funding, CRADAs being an important part of that. CRADAs have a unique advantage in that they can use part of our funding as a match for the industrial funding.

Cheryl Fragiadakis: Do you see a continuation of that basic philosophy? Do you have any order-of-magnitude expectation that a small percent of the research is likely to be partnership material, or is it more like half of the research?

William Millman: It is all good partnership material. I see this trend continuing, because when I look at the 2000 budget for Basic Energy Sciences, the research part is down $17.6 million, at least in the presidential submission. So we will continue to encourage collaborations.

Joseph Gordon, IBM: I just want to say that Henry Kohlbrand is exactly right in regard to the electronics products group. There is only one process for making semiconductors. Another structural difference from chemistry is that the major manufacturers of semiconductors depend on a large number of smaller manufacturers to provide key components. So it is not detrimental to them to have a program in which other people retain intellectual property. In fact, that is a key to making the entire system work. The roadmap, in fact, is key to coordinating all of these other components. Without the roadmap, none of the other companies will know which improvements to make, what to develop, and so forth. But I don't see why a technology roadmap could not be done in the chemical arena to identify key areas, for instance, desulfurization catalysts for energy conversion. If we see a regulation coming along, why can't industry get together and define what is required to meet it? There may be a problem in that the people who finally produce the gasoline also want to control the production of the catalyst. So, in this case, there is only one entity involved and, if there is only one entity, there is no room for a consortium or cooperation. There needs to be an analysis of what is actually being looked for in these cases.

9

Government–Industry Joint R&D Ventures: Bridging the Gap Between the Laboratory and the Marketplace

Lura J. Powell
National Institute of Standards and Technology

My discussion focuses on the Advanced Technology Program (ATP), as well as many of the other federal partnering activities that have not yet been mentioned at this workshop.

I want to start by emphasizing that the ATP is only one of the four major components of the National Institute of Standards and Technology (NIST). The core of NIST is its laboratories. NIST has a substantial chemistry presence throughout its laboratories, including the Chemical Science and Technology Laboratory, Materials Science Engineering Laboratory, Physics Laboratory, and Building and Fire Research Laboratory. The other two components of NIST are the Manufacturing Extension Program (MEP) discussed below and the Baldridge Quality Program.

As discussed by Christopher Hill, significant legislation was passed during the 1980s that encouraged partnering. Suddenly all of the federal labs were faced with a mandate to partner. Although this notably changed things for much of the federal government, at NIST it was business as usual. When NIST was created as the National Bureau of Standards (NBS) back in 1901, it was created with industry as the principal customer. It is the only federal laboratory with this mandate in the original legislation. When I started working for NBS in 1972, one of the first things I learned was that industry was my prime customer. From its earliest days, NBS had extensive collaborations with industry. And, in fact, before cooperative research and development agreements (CRADAs) linked industry and government in formal agreements, NBS actually had industrial research associate agreements. These agreements were executed when industry representatives would come to NBS to conduct research for a few months or more. For NBS, the biggest benefits of the technology transfer and partnership legislation came from the Freedom of Information Act exclusions, which facilitated joint research with industry.

THE NEED FOR PARTNERSHIPS

One of the messages I want to convey is how broadly partnerships have permeated the R&D enterprise. From my perspective, when looking across the R&D enterprise and at the varying R&D models, the critically important element is the leverage that partnerships can provide. Partnerships extend beyond industry and government, to include multiple companies (of varying sizes) working together with the academic community, the national laboratory system, and not-for-profit organizations.

Through this workshop, we have heard a lot about the R&D challenges of the 21st century. We know the effects that global competition has had on the U.S. R&D enterprise. We heard that we no longer have the industrial research environment of a few decades ago. Industrial R&D has become increasingly less long-term and much more directed. Industry has less flexibility to do the type of innovative research that resulted in many of the broadly applicable innovations we enjoy today. Thus, the real challenge for the 21st century is rapidly commercializing the fundamental innovations created in the universities and national labs. However, between the two ends of this spectrum there is a large gap, where an enormous amount of early-stage technology development work still needs to be done.

How do we bridge that gap? This is an important question, because there are few appropriate funding sources available. Although historically there has been a broad acceptance of federal funding for basic research, some policymakers question whether the government should fund industry to conduct R&D. From my perspective, I definitely see a role for government in filling this gap.

Many early-stage technologies are, by their nature, broadly enabling. Therefore, companies would not necessarily be able to capture all of the R&D benefits themselves if they were to invest in the research. In this competitive climate, with limited R&D budgets, companies focus strictly on what is going to benefit them and push them ahead of their competitors. Broadly enabling innovations are not likely to make the final cut of priority investments.

I am often told by industry that many of the ATP project ideas have been sitting on the back burner at the company for some time. Often, if a project was a priority of an individual researcher or group, it might have been bootstrapped along. This is exactly where ATP fits in, filling the gap. ATP is not the only government program that could fill this void, but it is the only one that sees filling this void as its primary mission.

I have been talking about the R&D process as a sort of continuum and have tried to capture this in Figure 9.1. ATP bridges that middle range of technology development. A number of ATP's projects still have some basic research components to complete when funding begins. Then the funding is halted at the point where the company has a prototype—ATP does not pay for product development or commercialization. The companies have to fund the last phase of manufacturing and commercialization on their own or seek funding from alternative sources.

BRIDGING THE GAP

I believe that government–industry partnerships are one way to bridge the gap. Not only can they help to address the funding issues that I mentioned above, but they can also tackle another R&D challenge: the increasing need for interdisciplinary expertise. Successful R&D requires broad partnerships and broad expertise. At ATP, for example, there has been a great increase in university involvement. We are also seeing an increase in the number of companies working together in partnerships. Often, a company will get an award from the ATP, and it will subcontract a portion of the R&D to another company, a university, or a national laboratory. Although ATP projects operate under a number of different models, each still needs a wealth of expertise to complete the research. One company may be able to do it alone, but to truly accelerate the development of new technologies, this collaboration is crucial.

Partnerships help to spread the risk, and ATP helps by buying down the risk. Through cost sharing, a company can reduce the risk of its investment in what it considers to be a not-yet-possible technology—those technologies that look promising, but might not ever emerge.

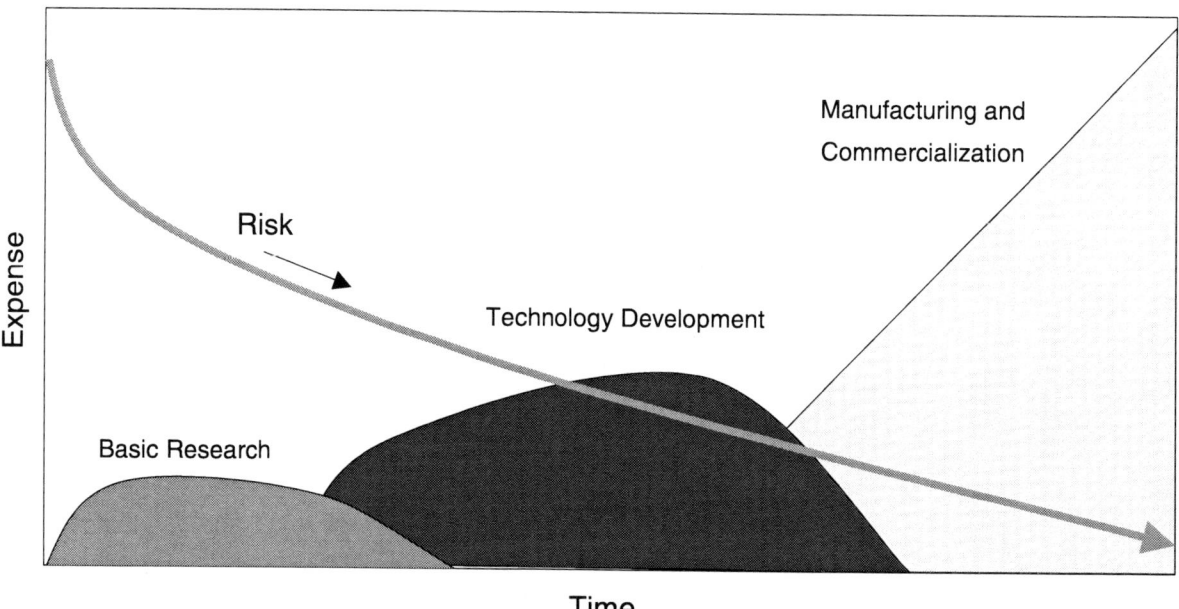

FIGURE 9.1   Characteristics of three phases of research and development.

From our perspective, the major benefit of cost sharing the research is that the government gets to leverage the enabling piece of the R&D—there has to be something in it for the taxpayer. One of ATP's key criteria is that the project will provide an economic benefit. Foremost for ATP is how the research will broadly benefit and bring economic value to the United States. Below are some examples of ATP projects that are doing just that.

Federal Partnering Programs

Before I turn my focus to ATP, let me first talk about some of the other federal partnering programs. There are a number of different partnering modes that are used at NIST. Other speakers at this workshop have discussed CRADAs; NIST does CRADAs somewhat differently. NIST has never had a "pot of money" that was set aside for partnering with industry. Instead, when companies have worked with NIST either individually or as part of a consortium, they have provided funding, a piece of equipment, or a researcher, while NIST provided the in-kind R&D expertise.

There are many other federal partnership programs besides CRADAs. The Dual-Use Applications Program (DUAP) is one example. It is administered by the Army, Navy, Air Force, DARPA, and the Director of Defense Research and Engineering. Christopher Hill discussed the rise and demise of the Technology Reinvestment Project (TRP). The DUAP is not a replacement for the TRP. Based on everything I have heard about it, it is much more applied than TRP. It began in 1997 with the mission to prototype and demonstrate new approaches for leveraging commercial research, technology, products, and processes into military systems. The concept is to take civilian technologies, products, and processes and turn them toward military use.

The Small Business Innovation Research (SBIR) program and the Small Business Technology Transfer Program (STTR) were developed during the 1980s when the focus was on collaboration. The SBIR program was created in 1982 by the Small Business Innovation Development Act to help small companies—defined as 500 or fewer employees—advance their innovative ideas and support their research that could lead to significant commercialization and public benefit if successful. To fund the SBIR program, a percentage of each agency's total extramural R&D dollars is set aside. Currently, roughly 2.5 percent of the extramural proceeds are used. In the case of the ATP, this amounts to about $5 million that goes to support NIST's SBIR program.

Somewhat more recently, the STTR came into existence to encourage technology transfer through cooperative research among small business concerns and research institutions. Currently, there are about five agencies that participate in the STTR program with the goal of promoting collaboration. The program requires a small business to partner with a not-for-profit organization, usually universities.

Over the years, NIST has tried some rather innovative partnership programs. The first was the Joint Institute Laboratory for Astrophysics (JILA), and the second, modeled after JILA, was the Center for Advanced Research in Biotechnology (CARB). JILA was created as a joint research initiative in the 1960s between NBS and the University of Colorado. It has developed into a wonderful collaboration, producing a lot of outstanding, award-winning, and internationally recognized research. Perhaps most importantly, they have incorporated industry collaboration. JILA is truly one of those innovative centers where a university and government partnership brings industry in as a third partner, therefore increasing its impact. Half of the principal investigators from JILA work for NIST, the other half are professors at the University of Colorado. JILA was so successful in Colorado that NIST decided to duplicate it in Maryland.

CARB was established in 1986. It was patterned directly on JILA; half of the principal investigators work for the University of Maryland Biotechnology Institute and the other half are staff of the NIST Biotechnology Division. It is truly a wonderful collaboration. In both of these organizations the collaborations work smoothly. In fact, the university and government scientists interact so freely, that I would challenge anyone to walk into CARB, talk to people, and then try to figure out which organization they belong to. At CARB, once again, industry researchers come in and take advantage of the facility. It is seamless. It provides a remarkable leverage for research dollars. When Rita Colwell was still head of the Maryland Biotechnology Institute and I was head of the NIST Biotechnology Division, we used to call CARB the research paradigm for the future. It has been a tremendous opportunity for the academic institutions, the agency, and industry.

In addition to these programs, there is also the NIST MEP. When the Competitiveness Act of 1988 was passed, the name of the National Bureau of Standards was changed to the National Institute of Standards and Technology, and the ATP and MEP were established as programs. MEP does not conduct R&D, but it is a major partnership program. The MEP is a nationwide network of more than 70 not-for-profit centers throughout the United States, with at least one in every state. MEP's mission is to disseminate manufacturing practices to small- and medium-sized manufacturers across the country. This is a similar concept to the agriculture extension service, disseminating technology broadly through a series of centers. In fact, we are currently exploring a MEP–ATP partnership. The ATP has concluded a number of projects, many of which developed manufacturing technologies. We are working with the MEP to develop a pilot program, which would take some of the manufacturing technologies developed under the ATP to see if they can be effectively diffused through the MEP centers to small manufacturing enterprises.

Partnerships with the Advanced Technology Program

Now, let me turn my focus back to the ATP. ATP's mission is to accelerate the development of innovative technologies through partnerships with the private sector. It is not involved with incremental technologies, but rather technologies that offer a leap forward—those that are path breaking. We accomplish this through partnerships with the private sector. Industry is the primary partner, but as I mentioned above, we have broad participation from other R&D communities as well. This being said, the bottom line for the ATP remains whether the research provides a broad national benefit. Therefore, each project is evaluated based on how the R&D will bring economic benefits to the United States.

ATP funds all areas of technology, and across the breadth of the technologies that we fund, there is a tremendous amount of chemistry. About half of the projects that were funded in microelectronics manufacturing last September had significant chemistry and materials science components. Some of the companies represented at this workshop are participating in ATP projects.

It is an exciting time at ATP. We are about to celebrate our tenth anniversary. To date, we have funded more than 400 projects with nearly 2,000 participants. We have over 1,000 participants, who participate as either principal awardees or joint venture partners. We also have nearly 1,000 subcontractors. It is easy to see by these numbers that we value and encourage partnerships under this program.

ATP has funded nearly $3 billion of R&D. A certain amount of that research is finished, but there is still roughly $2 billion ongoing. For each project, ATP provides half of the money, and half comes from industry. The research is conducted and managed by industry. More than 50 percent of our awardees are small businesses, and that number has been increasing. Although much of the research is done in industrial laboratories, a significant portion is done in universities.

Roughly 10 percent of the funding supports university research. There are 125 universities that participate as subcontractors or joint venture partners on ATP projects. We also have extensive national laboratory involvement, led by participation by Sandia and Oak Ridge. I have been encouraging the other national labs to do the same, because I believe that this type of collaboration will spark earlier-stage technology development. I want to see more universities and national labs linking proactively with companies at an early stage—when companies might be more inclined to form partnerships that help to put the United States in a more competitive position.

The 1999 ATP competition just closed. When the awards are announced in late summer, the ATP will be launching approximately $350 million in new R&D. Hopefully, next year this amount will be a little higher. So I encourage you to approach us with your ideas. One change we have made this year is the acceptance of preproposals year round. We hope this process will reduce the cost and risk to companies of applying to ATP.[1]

Measuring the success of the ATP is very important to us. ATP has an outstanding team of economists who are helping to build a strong story of ATP success. In fact, we just released a status report of the first 38 completed ATP projects, *Performance of Completed Projects: Status Report Number 1*.[2] The status report identified seven projects as potential home runs. When I identify a project as a home run, it must not only have a product in the marketplace, it must also offer big economic benefits. In fact, the benefits from three of these initial projects are projected to more than exceed the entire investment in ATP to date. Two of them are in the biological sciences, but one is in automotive

---

[1] Further details are available on the ATP Web site at <http://www.atp.nist.gov/>.

[2] William F. Long, *Performance of Completed Projects: Status Report Number 1*, NIST Special Publication 950-1 (Washington, D.C.: U.S. Government Printing Office, 1999).

process control. So it is not just the biological sciences that get the attention and have the potential for high benefits.

One small company, Engineering Animation, a software algorithm development company, came to us with only 20 employees in 1991. They now have over 900 employees and have developed software that allows medical students to study the human body in three dimensions. They then turned around and used that same base technology to create the animation for interactive games and movies such as *Small Soldiers* and *Animaniacs*. This was a huge home run, with broad applications.

Not all of the projects included in the status report were successes; 7 of these first 38 look like they are total failures. This is not a surprise because we fund high-risk R&D. In fact, we expect that some ATP projects will fail. When all is said and done, these projects are unlikely to contribute much to either the technical community or the economy. There are a variety of reasons for this. In some cases the projects did not finish the work. In other cases, management changed, the company went bankrupt, or the research did not yield the anticipated results.

Finally, there is a group of projects in the middle. Many of these projects could end up being real successes, but the jury is still out. Some of them have products in the market and others have linkages with other companies that could lead to applications; but the true success of the project has yet to be fully revealed. These fall into a "watch and see" category. With a program like the ATP, you expect to see this mixture of projects with moderate benefits as well as projects that are going to be real home runs. We do not expect them all to be home runs; but only by funding a large number of the truly innovative ideas will we ever see the home runs.

When thinking about potential ATP projects, think about what you will not do on your own. ATP is looking for are these really innovative, high-risk ideas. If it looks like the project could receive funding elsewhere, then we are not going to fund it.

We are getting a lot of positive feedback from our partners. "[Our] members regard an ATP award as the gold medal in the research Olympics." This quote from the Ohio Aerospace Institute (OAI) is not only wonderful because of how it captures the prestige felt by ATP participants, but also because it allows me to highlight the fact that we have a number of not-for-profit companies that successfully lead joint ventures. OAI is one of them and so is the National Center for Manufacturing Sciences, which Christopher Hill mentioned in his presentation.

Having not-for-profit companies leading joint ventures works very well. Not only do they know how to attract different partners, but they also know how to keep them together and find new partners when necessary.

ATP stimulates collaborations. We have conducted many surveys with ATP participants over the years. A majority of participants have told us that collaborations helped them achieve their project goals. That includes collaboration with either a subcontractor or joint venture partner. We also hear that ATP is greatly responsible for stimulating these collaborations, a fact that makes us very happy. Even at the height of the ATP debate in the mid-1990s, Congress and the General Accounting Office gave us credit for our role in stimulating collaborations and joint ventures.

One of the main benefits highlighted by our companies is the ability to complete complex projects by obtaining the necessary interdisciplinary R&D expertise. These research relationships are credited with stimulating the creative thinking necessary to reach project goals. Also highlighted was the acceleration of entry into the marketplace facilitated by collaboration. Most of our companies have seen a two-year or more acceleration in the marketplace, essentially saving time by dedicating a significant number of people with varying skills to work on the project. Collaboration has also helped to better identify customers' needs. Because these partnerships often include companies from a wide range of industries, it is almost like having an entrance to many different markets initially.

A very exciting outcome emerging from these partnerships is the future collaborations that they are stimulating. Individuals who have collaborated for the very first time under the ATP continue to do so and have a greater tendency to collaborate in the future. We are encouraging this collaboration and would like to see more of it in our projects.

Last year at our first national meeting we made stimulating collaborations a goal. We brought together potential proposers and past awardees to attend our technical and proposal workshops. We also featured an exhibit in which ATP participants could display and talk about their technologies. Following the meeting, the exhibitors shared with me how the conference had provided them with the opportunity to sell to people they never would have marketed to before. This type of networking really excited us. So at the 1999 national meeting, which will be held in San Jose in mid-November, we are doing everything on a much grander scale, including the exhibits.

ATP has a long history in holding technical workshops, particularly in the days of focused programs. Many chemists and chemical engineers have attended workshops in the chemical catalysis and separations areas. One of the first workshops that I attended as the ATP director was in the composites field. At the workshop, two of the ATP awardees that made presentations were Strongwell Corporation and Ebert Composites. Ebert Composites has developed a real innovation for making very large transmission towers out of composite materials, and Strongwell has developed technologies to make strong and durable composite bridge supports for transportation infrastructure. These two companies met for the first time at the meeting. Since that first meeting, they have formed a whole new company together. It is these and other collaborations that we try to foster, even beyond the ATP.

Although we do not have many completed projects from the chemistry field, we do have a significant number that are under way. One is a breakthrough process for direct oxidation of propylene to propylene oxide, led by the Dow Chemical Company. Another involves continuous biocatalytic systems for production of chemicals for renewable resources, involving collaboration among Genencor, Eastman Chemical, and Argonne National Laboratory. Yet another involves biosynthesis of monomers, led by General Electric. These are just a few of the projects; there are many more and all of them can be found on our Web site.

A recently completed project is one that involves Air Products and Chemicals, Inc. It is a joint venture with Toromont Process Systems, Inc., which developed a closed-cycle refrigeration technology based on air. It is a novel system that has gone through a successful test at Kodak and now looks as if it is poised for positive sales and broad impacts. Although the results are so recent that we have not published a success story yet, we are in the process of putting one together.

We have a lot of success stories like this one. I have a couple of other successes that I want to go over quickly. Although they are not necessarily from the chemical industry, I think they will give you an idea of some of the things that we fund and look for.

Back in 1991, a consortium of the six top printed wiring board companies along with Sandia approached ATP for funds, mainly because the U.S. printed wiring board industry was in danger of disappearing. It was moving offshore, and the United States was in danger of losing this critical technology. At the time, the U.S. market share had dropped from 40 percent down into the low twenties. These companies partnered to tackle a suite of 64 projects. Many of these projects involved chemistry: materials science, resins, etc. However, a large number centered on improving manufacturing capabilities. The collaboration in this project was highly beneficial. The National Center for Manufacturing Sciences (NCMS) has estimated that without the ATP funding, about half of these 64 projects would never have been done and the other half would have been bootstrapped along and done at a slower pace. By doing the R&D jointly, the industry saved more than $35 million in research dollars. Now the U.S. printed wiring board industry is making a comeback, with its market share rising from the low twenties back up into the thirties. NCMS, which was responsible for pulling the industry together, has given ATP, and the researchers involved, the credit for saving the 200,000 jobs in this country that would have been lost if the printed wiring board industry had moved offshore. Most importantly, this group of six companies, through government cost sharing, brought an industry back.

The final success I wanted to share with you is one of the home runs from the status report mentioned above. It was done by a group of small- and medium-sized supplier chain manufacturers called the Auto Body Consortium. The Auto Body Consortium also partnered with two universities, University of Michigan and Wayne State University, and two of the big three auto manufacturers, General Motors and Chrysler. As it turns out, the "brains" of the project was a mechanical engineering professor at the University of Michigan. He had the original inspiration, grabbed some partners, and said, "I've got this neat idea." What followed stemmed from this "neat idea."

For this discussion, this project offers three central messages. First of all, the technology was very exciting and leading edge. When the consortium first started this project in the early 1990s, the U.S. automobile manufacturing industry was drifting further and further behind its competitors with respect to the dimensional fit and finish of automobile parts. To combat this problem, the Auto Body Consortium developed a novel process navigational system and series of sensing technologies. This system takes the automobile from design all the way through manufacture, allowing the manufacturers to get the fit and finish down to less than 2 millimeters—a significant advance. Furthermore, the technology will also help them pursue even tighter tolerances in the future.

A second point, particularly of interest to the ATP, is that the technology is broadly enabling. The same technology used for the automobile industry is being now diffused to the furniture and appliance industries.

The last point I want to make is for those small- and medium-sized companies that have trouble with 50–50 cost share. General Motors and Chrysler basically paid the full freight; they did not take any ATP money to cover their costs, and they covered the majority of cost sharing for the project. Now they are reaping the benefits. This technology is in more than 20 plants across the United States. In fact, the Chrysler Concord that was redesigned over a year ago incorporates this technology.

I want to close with a quote that is about five years old, but still extremely valid today:

> The United States has unparalleled resources of science and technology. Its industrial research capability, universities, nonprofit research institutions, and federal laboratories are great national treasures. But in a time of severe financial constraints and heightened international competition, the Nation must maximize its return on those assets. . . . The time is ripe for bold steps to capitalize on the promise of partnership.[3]

The United States has tremendous resources for science and technology, and, sadly, our financial constraints are not going to get any easier, so the time is ripe for partnering. I really believe partnerships are the wave of the future. I encourage each of you to partner, and where it makes sense, partner with the government, with the ATP.

DISCUSSION

Fritz Kokesh, Massachusetts Institute of Technology: With all of the success stories that you describe, is there not enough information available to put to rest the corporate welfare argument?

---

[3] State-Federal Technology Partnership Task Force, *Recommendations of Final Report* (Washington, D.C.: NSTC, 1995).

Lura Powell: It has been put to rest to a great extent. About a month ago, under the auspices of the National Research Council, we had a meeting specifically on the ATP. Based on the comments, I was pleased, as ATP director, to hear that the controversy about whether ATP should exist or not has basically disappeared. Instead, the panel discussion focused on how big the program should become, how fast it should grow—budget caps being what they are.

Christopher Hill, George Mason University: I agree completely.

Fritz Kokesh: Just to follow up, I hope the data you speak of will be publicized because, in contrast to some other things that we have talked about at this workshop, the ATP has had enormous success. At the same time, you indicate that there is a change in the wind. Do you understand well enough the reasons why the ATP has been successful to predict the effects of the change? Have you started to tweak the system? Do you understand why the program has allowed groups to work together in the ways that they have?

Lura Powell: Yes, we are beginning to, although I am unsure that we understand all of it. But we practice program evaluation as an integral feature of the ATP, and that is adding in a systematic way to our understanding of what, why, and how it is working. One of the things that has been attractive to certain companies about partnering under a federal government program is that it makes them more comfortable taking the step to collaborate without intense fear that the collaboration will automatically be viewed as a violation of antitrust laws. At the outset of the program, it was generally expected that the joint ventures would be where the collaborations took place, as opposed to the single-company awards. But we have found that extensive collaborating with other companies, universities, or others occurs in most of the ATP projects, including the so-called "single-company projects." There are a number of factors that account for this, including the fact that most of the technology development projects we fund are complex, requiring state-of-the-art multidisciplinary expertise that few single companies possess in the entirety. Another driver of collaboration is the fact that our selection criteria encourage up-front integration across science and business goals. This causes the market-oriented companies to reach out to strong R&D partners and vice versa in order to show strength across the board. And when companies are developing technologies that will benefit their entire industry, and other industries, they often seek partners to share the costs because they will all share the benefits.

We also try to be very user friendly. There are many people here at this workshop who have ATP awards. Often when I give talks, I encourage ATP participants to share their experience of working with the ATP. I do this in part because we are constantly trying to figure out how the ATP can better serve industry. We have also taken great pains to see how we can be more user friendly to universities and have made changes to encourage more university collaboration. In fact, 137 universities are currently participating in 234 ATP projects.

The ATP is a government program that was created, in part, to promote collaboration and partnership. And I believe that it is this core characteristic that has led to our success. As to why we are achieving big benefits, I think it is attributable to our focus on the potential economic benefit and understanding why these people need the ATP. We only fund the development of new technologies that are expected to have widespread beneficial effects extending well beyond the innovators. And we only fund where the ATP can be expected to make a big difference. We recently restructured our selection criteria to clarify that.

Ven Narayanan, National Institutes of Health: I must congratulate you on your success story of building this partnership, particularly the type of research leading to products in the marketplace. Are there any restrictions on some of your payoff dollars returning back to support other new projects?

Lura Powell: You mean recoupment?

Ven Narayanan: Yes, so that you can fund more such projects. Any prohibitions with that?

Lura Powell: Actually what we look at as the payback mechanism is job creation and company growth and the economic benefits that go to others outside the ATP award recipients—what economists call the "spillover benefits." Basically there is a much bigger impact in the longer run as higher-quality jobs are created, industry sectors are strengthened by technology-induced productivity gains, and the tax base grows. But beyond the benefits that show up in standard economic measures of national income, there are benefits that come from improved quality of life—health and safety benefits, environmental benefits, and other advantages that come from a higher standard of living through innovation.

Ven Narayanan: I asked the question because you said that the dollar amounts are limited.

Lura Powell: I am not sure whether you can actually do recoupment on early-stage research. It doesn't seem to make sense for our type of projects. By the time the research moves forward to the point where there's a commercial product, there have been so many other funding sources involved that it would be almost impossible to figure out ATP's piece of the technology to recoup against. It would be a bit of a morass and might cost more than you could get back.

Ven Narayanan: I wanted to share some information as well. The National Institutes of Health is a very discipline-oriented, grantee-oriented type of institution. We face this problem with partnering—talents across the universities and industries in different geographic locations.

We started the national cooperative drug discovery groups in natural products and we now have groups in chemistry and mechanism-based approaches where we partnered with individuals utilizing some of their specific talents to solve multidisciplinary problems. The government funds the project at a similar level. We have about $8 million to $10 million allocated to this area.

The program has been successful and has resulted in the development of at least three novel cancer drugs.

10

Factors Contributing to the Success of the Partnership for a New Generation of Vehicles

Christine S. Sloane
General Motors Corporation

The Partnership for a New Generation of Vehicles (PNGV) is a cooperative undertaking to conduct R&D for a significant advancement in the energy efficiency of passenger transportation in the United States. The partnership is focused on a number of challenges that require basic research into the chemistry and physics of new materials, which are coordinated toward the achievement of quantitative technical performance goals that are linked through system analysis models ultimately to the fuel efficiency of vehicles.

Because the partnership targets and coordinates this broad spectrum of R&D projects under the cooperative leadership of the U.S. government and the three principal domestic vehicle manufacturers, it offers a model for other cooperative research undertakings to consider. The objective of these remarks is to convey an industry-based perspective about six selected features of this partnership that have been essential factors in its success. First, however, let us review some of the defining elements of the partnership to establish a common basis of reference.

BACKGROUND: PROGRAM DESCRIPTION

The PNGV was established in the fall of 1993 by the then newly elected Clinton administration and the three principal domestic automotive manufacturers. It addresses three goals: the development of technologies for highly efficient five-passenger vehicles, for their reliable and cost-effective manufacture, and for the spin-off transition of these technologies into passenger vehicles as soon as commercially feasible. All aspects of energy efficiency were targeted: powertrain, aerodynamics, mass, rolling resistance, and regenerative energy capture and reuse. Several scenarios of design targets, such as a scenario illustrated in Figure 10.1, are evaluated in vehicle simulation models to determine whether they are consistent with the desired threefold improvement in overall vehicle fuel efficiency. The challenge has been to identify scenarios that have the potential of being realized within the PNGV time frame under an aggressive development program.

The schedule for the program requires considerable discipline to identify research requirements to support candidate technologies, to monitor progress relevant to automotive requirements, and to focus

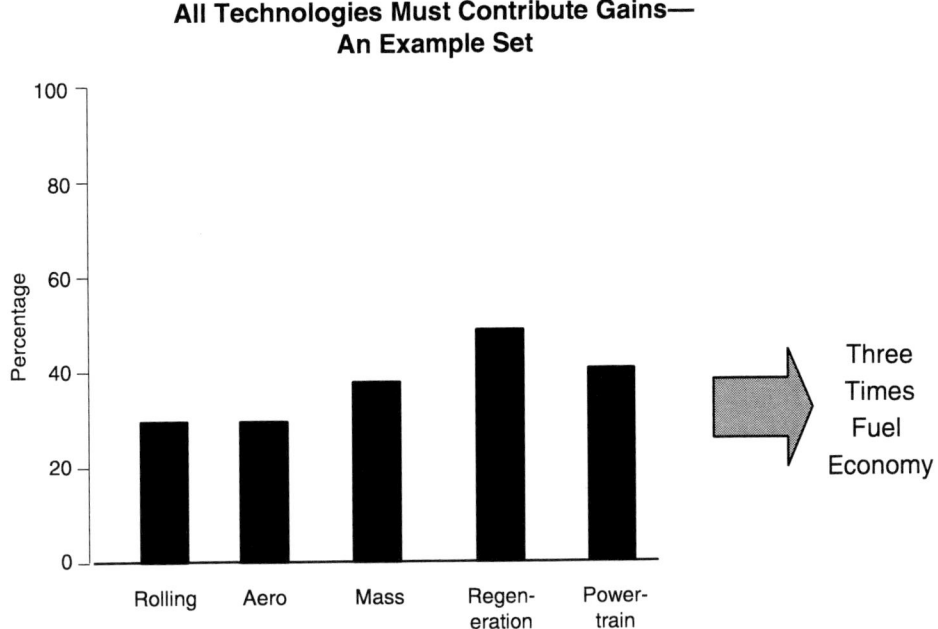

FIGURE 10.1   Reaching a threefold increase in fuel economy.

resources consistent with the requirement to demonstrate proof of technical concept at the vehicle level in calendar year 2000. Toward that end, technology selections scheduled for calendar year 1997 were accomplished and focused attention on lightweight materials, four-stroke direct injection engines, hybrid electric drivetrains, and fuel cell systems (see Figure 10.2).

Proof-of-technical-concept vehicles will demonstrate the fuel economy attainable in fully realistic five-passenger vehicle designs during calendar year 2000. The concept vehicles are deliverables for the industry that is responsible. The government is responsible for support of the high-risk R&D needed to create the knowledge base that underpins the technology development. The next stage of the program will focus on improved vehicle utility and reduced cost as the program drives toward production feasibility. The many needs for new understandings in chemistry and physics will underlie the manufacturing challenges in that focus period of the program (see Figure 10.3).

ELEMENTS OF THE PROGRAM CRITICAL TO ITS SUCCESS

Of the many critical features of this program, six can be highlighted as essential to its success from an industry perspective:

- Compelling significance: societal benefit and technical challenge,
- Commitment,
- Cognizance of market forces,
- Reinvention of relationships,
- Diverse resource base, and
- Accountability.

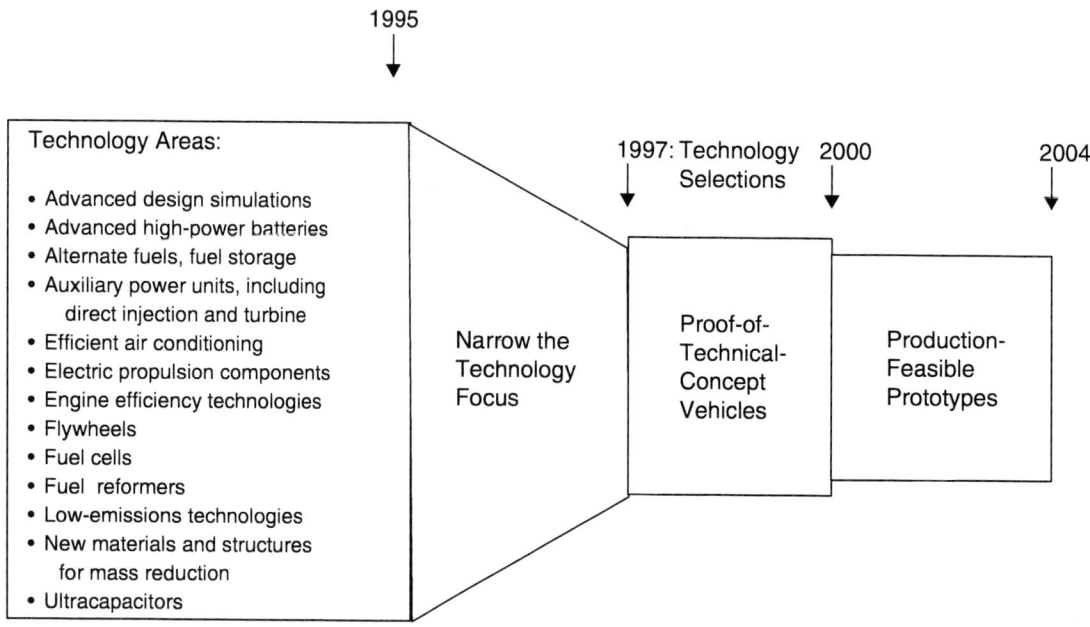

FIGURE 10.2   PNGV schedule. The plan: Invent on schedule.

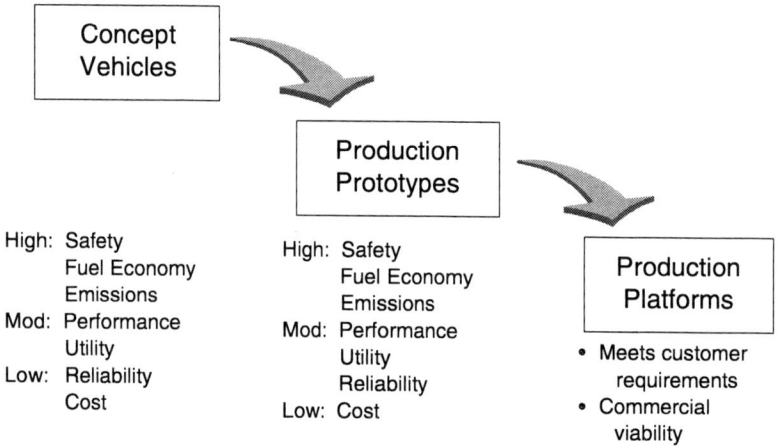

FIGURE 10.3   Evolving priorities.

This listing is not intended to be comprehensive, but to provide focus points for discussion.

A *compellingly significant societal benefit* is essential to build the breadth of support required to attract inspired technical thinking and to maintain financial support over the full duration of the program schedule. From the government's perspective, the PNGV program addresses several pressing national objectives: U.S. energy security, global warming, balance of trade (oil and auto imports), and economic competitiveness. From industry's perspective, the PNGV program provides, in addition, a U.S. demonstration of nonadversarial government–industry problem solving and the leveraging of resources toward U.S. transportation objectives. Nonadversarial government–industry problem solving and leveraging of resources have prospered in the health, agricultural, and defense sectors. PNGV offers a complementary model to improve the quality of life and meet significant national objectives in transportation.

A *compelling technical challenge* has historically attracted the best technical talent to think of novel solutions and accept the risk of pursuing them. The challenge of designing a car that gets up to 80 miles per gallon with comfort, performance, and utility for five passengers while maintaining compliance with safety and emissions requirements has demanded novel and cooperative thinking for every component and subsystem and every manufacturing process to lay the foundation for the 21st century.

Experts stimulated by seemingly impossible technical targets have risen to the challenge to find solutions that have built the probability of success from near zero at the beginning of the program (when no solutions were evident) to today's favorable expectation for the proof-of-technical-concept vehicle demonstrations in calendar year 2000. The subsequent phase of the program, the production-feasible prototypes in calendar year 2004, is likely to present even greater technical challenges.

Commitment at the highest responsible levels of the participating organizations has been critical to maintain resources and a sense of significance. In an environment of fixed government research funding, support of good new ideas and pursuit of worthy new objectives requires that funds be diverted from programs already under way. In industry, competitive requirements to cut costs and meet current market challenges result in continuing pressure to divert funding from ongoing development programs. A very public and personal commitment to the goals of the PNGV program by the government leaders in both the administration and Congress, and in industry, has been essential to maintain the focus of resources on the program objectives over the past five-year period.

Cognizance of market forces is essential for significant and long-term industry participation. American market forces in the personal transportation sector are dominated ultimately by preferences of the U.S. customer. As evidenced in Box 10.1, customer preferences for fuel economy reflect the price of

---

BOX 10.1
Purchase Priorities for U.S. Car Buyers in 1980 and 1994

Low oil prices have reduced market forces to support private investment in energy efficiency.

1980 Purchase Priorities
1. Durability and reliability
2. **Fuel economy**
3. Well-made vehicle
4. Value
5. Price or deal
6. Dealer service
7. Warranty coverage
8. Riding comfort
9. Safety features
10. Cost of service

1994 Purchase Priorities
1. Durability and reliability
2. Well-made vehicle
3. Ease of handling
4. Safety features
5. Riding comfort
6. Value
7. Price or deal
8. Manufacturer's reputation
9. Run to drive
10. Exterior styling . . .
15. **Fuel economy**

gasoline, which is currently at a historical low as a percent of average income. As a result, current market forces do not support responsible investment of industrial investment in fuel economy. Two offsetting factors are critical to this program. First is the awareness that current market forces can change, and investments in technologies for improved fuel economy provide security against that shift. Second, and of primary importance, is the leveraging of the investment risk by government funding of targeted R&D to offset the market deterrents caused by individual purchase choices in favor of the societal benefit.

The *reinvention of relationships* has characterized PNGV in four respects. First, the competing partners (the domestic manufacturers) agreed to fundamental guiding principles for the PNGV program: (a) the prime objectives are the achievement of technical goals and credibility in progress toward societal goals and national competitiveness; (b) credibility requires sustained commitment and visible progress; and (c) collaboration will be the preferred approach whenever feasible. Second, the working relationship between industry and government participants was defined. Government and industry technical communities committed to jointly defined goals and a vision of success, which developed a common technical strategy. In the implementation of the joint technical strategy, the government role focused on provision of R&D resources, convening power (national laboratories and government suppliers), and coordination of regulatory and infrastructure issues. The industry role focused on provision of R&D resources, convening power (industrial supply base), and coordination of market requirements. The joint technical strategy was developed in support of vehicle-level requirements by identifying and prioritizing technical opportunities; setting performance targets and milestones for subsystems, components and materials; and identifying resource responsibilities (see Figure 10.4).

Next, relationships between the principal automotive manufacturers and their supply base shifted to include far-term technology development targets and the rapid deployment of evolutionary applications. The domestic manufacturers provided encouraging evidence of their commitment to pursue commercial product opportunities. Suppliers were encouraged to cooperate to leverage their capability for more rapid technical progress. Last, relationships with universities were shifted by linking research requirements to performance targets and program schedules, and by coordinating industrial research objectives with government objectives and programs.

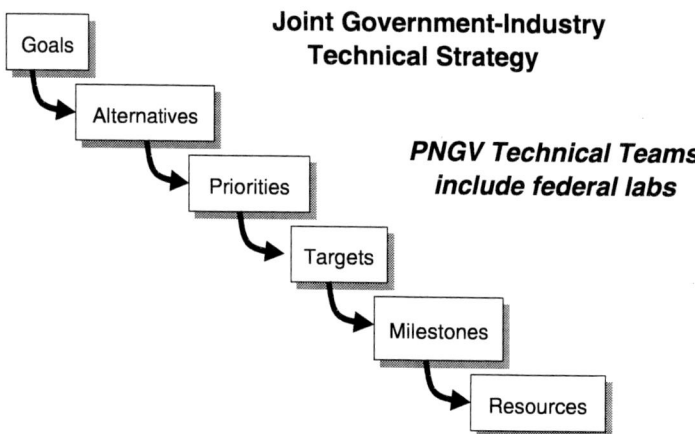

FIGURE 10.4   PNGV strategy.

A *diverse resource base* beyond the traditional automotive technical community has been required to leverage rapid advancements in computing and electronics for hybrid electric drivetrain development, to tap opportunities for fuel cell systems, and to tackle the breadth of materials requirements to meet the PNGV goals. Technical talent has been required with basic, developmental, and applied expertise. Matching the breadth of technical talent has been a broad funding base including the federal government, the automotive manufacturers (collectively and individually), and the industrial supply base.

Key issues in maintaining stability in government resources have been a commitment by otherwise divergent political interests to the goals of the program and its evidence of progress and careful consideration of the alignment of individual government agencies' missions with PNGV technical objectives. Early concerns about corporate welfare were resolved by recognition that little of the government funding (less than 1 percent in 1999) passes through the automotive manufacturers, that the industry contribution exceeds the government contribution by a considerable margin, and that the government role serves to offset adverse market forces to encourage development of technologies toward national energy efficiency. A continuing issue remains with regard to government cost-share requirements that discourage participation by small innovative companies and noncommercial entities such as universities and contract research organizations.

Accountability is, perhaps, a unique feature of the PNGV. Accountability consists of three elements: hardware deliverables, visible milestones, and external technical audit. The hardware deliverables, which are obligations of the industry partners, consist of proof-of-technical-concept vehicles to be delivered in calendar year 2000 and production-feasible prototypes to be delivered in calendar year 2004.

Visible milestones have been produced each year during the program. They consist of measurable evidence of achievement of subsystem performance targets and intermediate concept vehicles presented annually at the North American International Auto Show. Achievement of visible milestones has been critical to sustaining the technical credibility of the program.

Finally, the program is subjected to extensive external technical audit. The formal element of audit occurs under the direction of the National Academies of Science and Engineering, which conduct an annual technical review and publish an annual report.[1] The review has provided constructive insight and is generally quite supportive of the program priorities, targets, and progress. The informal element of technical audit occurs through the technical new media, which reports frequently and aggressively on the PNGV program.

All of these elements of accountability have functioned to ensure that the program maintains diligence in continually evaluating technical opportunities, funding priorities, measures of progress, and relevance toward the ultimate objectives.

DISCUSSION

Joseph Cecchi, University of New Mexico: The issue of globalization has come up a few times, and here is an example in which this impacted in the middle of your program. Can you comment on how the Daimler Benz acquisition of Chrysler affected their participation, and in particular is all of the Mercedes Benz research in this topic part of the group now?

---

[1] For the latest report, see National Research Council, Standing Committee to Review the Research Program of the Partnership for a New Generation of Vehicles, *Review of the Research Program of the Partnership for a New Generation of Vehicles: Fifth Report* (Washington, D.C.: National Academy Press, 1999).

Christine Sloane: The United States Council for Automotive Research (USCAR), the organization through which collaborative R&D is undertaken within the domestic automobile manufacturers, is the organization through which PNGV is coordinated with the federal government. USCAR reaffirmed its mission and membership criteria. Those criteria include significant basic R&D facilities and engineering activity within the United States. Another criterion is that the full technical resources of the members are brought to bear on collaborative projects.

Joseph Cecchi: So that was a good deal.

Christine Sloane: If you mean that the R&D base is broader, yes.

Christopher Hill, George Mason University: I want to ask two quick questions about the political objectives of the program and whether they have been achieved. And the questions are impertinent, but they are highly relevant. In the year 2000 Democratic primary in Michigan, will the United Auto Workers support Bradley or Gore?

Christine Sloane: I don't know.

Christopher Hill: It's a critical political outcome because United Auto Workers' support for candidate Clinton was one factor that drove the program originally.

Christine Sloane: Let me add a personal perspective on your assertion. In 1994, the majority party in the Congress changed and advocated different political objectives than the administration. There was considerable scrutiny to eliminate "corporate welfare" programs, and the PNGV program was reviewed extensively by the new majority party in that light. The majority party, through its actions in the appropriations process, has supported the PNGV program at roughly constant levels over the past five years. So while Mr. Gore was very involved in the initiation of the program, its supporters crossed party lines both at the outset and today. The reason, of course, is that the societal benefits for the country override party allegiances.

Christopher Hill: That's an important observation. The second question, closely related, is that this program has managed to forestall any increase in Corporate Average Fuel Economy (CAFE) standards over seven years, but what it hasn't forestalled is the radical shift in market share from passenger autos to trucks and sport utility vehicles (SUVs). We are now beginning to see pressure to impose on them some of the performance standards that passenger vehicles have been under. How will that affect this program?

Christine Sloane: The PNGV program charter refers to passenger vehicles, not cars. A primary milestone of the program, however, is the delivery by the automobile manufacturers of proof-of-concept cars that provide up to 80 miles-per-gallon fuel economy, safety, emissions, and utility comparable to midsize passenger cars. So in that sense, the program's focus is stronger on cars than on trucks. However, virtually everything that we are developing collectively—for lightweight structures, compression ignition direct injection-based powertrains, emissions control, electric drive componentry, fuel cell componentry—can be applied to SUVs and trucks. Indeed, all of us are seriously looking at applying these technologies across the product line, definitely including SUVs and trucks. Indeed, if you look at this year's auto shows, you will see fuel cell power systems in SUV-concept vehicles.

Ashok Dhingra, Dupont: What part of your new concept car in 2000 will involve materials innovation and what will involve the innovation in energy such as fuel cells and other competing technologies?

Christine Sloane: We cannot approach the threefold fuel economy without employing the most advanced technologies for powertrain, reduced vehicle weight, reduced auxiliary power losses, and reduced aerodynamic and rolling losses. In the drive for reduced weight, two primary factors influence material usage: strength to maintain excellent safety performance and cost to provide customer affordability. The year-2000 vehicles will be revealed in January with full descriptions of material use. Early indicators suggest that there will be at least two aluminum-intensive designs.

We expect that the three companies will produce three different realizations of lightweight vehicle design. Two will be aluminum-intensive, that is, involving aluminum structure. However, all three manufacturers will use composites extensively where weight savings can be achieved consistent with reasonable cost potential.

Ashok Dhingra: And the composite materials?

Christine Sloane: The vehicles have not been announced yet, so a definitive statement of materials usage today would be premature. But surely, just as today's vehicles are not made of one material, these concept vehicles will use composites extensively. By the way, it can be misleading to cite material usage in a vehicle in terms of the percentage of the vehicle weight. For example, if half of the material in a car is replaced by new materials that weigh half as much as the old materials, the weight percentage of the new materials will only be 33 percent, not 50 percent, even though the lighter materials have replaced half of the car.

James Desveaux, University of California, Los Angeles: Do you think that your experience with USCAR, or for that matter with any other significant collaboration, has fundamentally altered the climate of trust between General Motors and the government? You made a point early in your talk that government, whether at the state or federal level, often sets technological standards for you to meet and only later realizes that for technological reasons, they have set a threshold that is too high and is one that you cannot meet. Government understands that they need to throttle back on requirements, while industry has spent millions or tens of millions in an effort to meet those standards, and they see little or no reward for the effort.

But from the government's perspective, there is a history of disingenuousness on the part of the auto industry. Here I am thinking of the CAFE standards, and the claims by industry that these and other Environmental Protection Agency requirements were impossible to meet, which were proved in many instances to not be the case. So my general question is, have some of these mismatches and problems in perception and trust between you and government bureaucracy gotten better as a result of more collaboration in R&D?

Christine Sloane: Trust is hard to measure; it's an outgrowth of in-depth understanding, evidence of actions, and mutual experience in working problems. There is no question that the PNGV program has afforded the federal government insight into the development of technologies within the automotive industry and the opportunity to validate its assessment of technological challenges. Indeed, the technical challenges are fairly clearly identified, and the state of progress in the collaborative projects to surmount them is evident to the participating government scientists. So the government agencies that are working with us have a sense of reality about what is technically feasible. They are entrained in the challenge too. They don't always have to take our word for it. In a sense, that alleviates the reliance on trust, or from another perspective, it buttresses trust. One caveat is that the Department of Energy is the primary government agency that is engaged with us in solving the technical challenges, so other government agencies have a less direct basis for understanding.

Beyond the technology projects, there is considerable evidence that PNGV is creating a push to develop advanced technologies for enhanced fuel economy. Early PNGV activity appears to have spurred early activity in JCAR and EUCAR (corresponding Japanese and European car programs), which in turn create a competitive environment for fuel efficiency. As a result, Toyota has announced the introduction of a moderate-sized car with a hybrid electric drivetrain for the United States in 2000, and General Motors has announced a hybrid electric truck, just to name a couple. So the spin-offs are already beginning. The PNGV focus, however, is to push the development to an even higher level of fuel efficiency.

Appendixes

APPENDIX A

List of Workshop Participants

Richard C. Alkire, University of Illinois at Urbana-Champaign
Michael W. Babich, Florida Institute of Technology
J.P. Baselt, Dechema E.V.
Kristin Bowman-James, University of Kansas
Michelle V. Buchanan, Oak Ridge National Laboratory
R. Morris Bullock, Brookhaven National Laboratory
Donald M. Burland, National Science Foundation
Glennia R. Campbell, Stanford University
James O. Casey, Eastman Chemical Company
Joseph L. Cecchi, University of New Mexico
Michael A. Chartock, Lawrence Berkeley National Laboratory
Khirod C. Das, Commonwealth of Virginia
Ronald A. Demarco, Office of Naval Research
James A. Desveaux, University of California, Los Angeles
Ashok K. Dhingra, E.I. du Pont de Nemours & Company
Louise B. Dunlap, Oak Ridge National Laboratory
John G. Ekerdt, University of Texas
D. Fennell Evans, University of Minnesota
Frank J. Feher, University of California, Irvine
Cheryl A. Fragiadakis, Lawrence Berkeley National Laboratory
Benjamin H. Franklin, Dow Corning Corporation
Jean H. Futrell, Pacific Northwest National Laboratory
Joseph G. Gordon II, IBM Research Division
Harold C. Graboske, Jr., Lawrence Livermore National Laboratory
Randolph J. Guschl, E.I. du Pont de Nemours & Company
Bruce J. Harrer, Pacific Northwest National Laboratory
Joe W. Hightower, Rice University

Christopher T. Hill, George Mason University
Nancy B. Jackson, Sandia National Laboratories
Andrew Kaldor, Exxon Research and Development
David S. King, National Institute of Standards and Technology
Kenneth Kirk, National Institutes of Health
William Koch, National Institute of Standards and Technology
Henry T. (Hank) Kohlbrand, Dow Chemical Company
Fritz Kokesh, Massachusetts Institute of Technology
Todd R. La Porte, University of California, Berkeley
Dennis L. Lichtenberger, University of Arizona
Robert Y. Lochhead, University of Southern Mississippi
Thomas Malloy, UOP/Growth Technology Center
Thomas A. Manuel, Council for Chemical Research, Inc.
Michael Marron, Office of Naval Research
Donald E. McLemore, Raychem Corporation
Claude F. Meares, University of California, Davis
William S. Millman, U.S. Department of Energy
David C. Mowery, University of California, Berkeley
Ven L. Narayanan, National Institutes of Health
Arthur Nozik, National Renewable Energy Lab
Janet G. Osteryoung, National Science Foundation
Lura J. Powell, National Institute of Standards and Technology
James B. Roberto, Oak Ridge National Laboratory
Michael E. Rogers, National Institutes of Health
David G. Schetter, University of California, Irvine
James N. Seiber, U.S. Department of Agriculture
Debonny L. Shoaf, Idaho National Engineering and Environmental Laboratory
Christine S. Sloane, General Motors Corporation
Vedene H. Smith, Queens University
Alan P. Sylwester, Sandia National Laboratories
John C. Tao, Air Products and Chemicals, Inc.
Matthew V. Tirrell, University of Minnesota
Laren M. Tolbert, Georgia Institute of Technology
David L. Venezky, Naval Research Laboratory
Francis A. Via, General Electric
William A. Wakeham, Imperial College of Science, Technology, and Medicine
F. Ben Wampler, Los Alamos National Laboratory
Henry F. (Hank) Whalen, Jr., PQ Corporation
Ralph E. White, University of South Carolina
W.B. Whiting, University of Nevada
Elizabeth K. Wilson, Chemical & Engineering News
Patrick H. Windham, Windham Consulting

Staff

David A. Grannis
Ruth McDiarmid
Sybil A. Paige
Douglas J. Raber

APPENDIX B

Biographical Sketches of Workshop Speakers

D. Fennell Evans is the director of the Center for Interfacial Engineering and professor of chemical engineering and materials science at the University of Minnesota. He is the author of more than 180 publications on self-assembly processes in water and nonaqueous solvents, microemulsions, diffusion in liquids and micellar solutions, and characterization of surfaces using scanning probe techniques. He has published two textbooks, *The Colloidal Domain* and *The Fundamentals of Interfacial Engineering*.

Christopher T. Hill is vice provost for research and professor of public policy and technology at George Mason University in Fairfax, Virginia. He teaches courses in science and technology policy, industrial R&D, and the logic of policy inquiry. His current research is on the impacts of cooperative R&D on participating firms. Before joining George Mason University in 1994, Dr. Hill was at the RAND Critical Technologies Institute in Washington, D.C. He has also held positions at the National Academy of Engineering, the Congressional Research Service, Massachusetts Institute of Technology, Washington University in St. Louis, the Office of Technology Assessment, and the UniRoyal Corporation. He has a B.S. degree from the Illinois Institute of Technology and M.S. and Ph.D. degrees from the University of Wisconsin, all in chemical engineering. He is a registered professional engineer in the State of Missouri, and he is a consultant to the New Energy Development Organization of Japan and to the Procter & Gamble Company.

Nancy B. Jackson is a principal member of the technical staff and chair of the Catalysis Steering Committee at Sandia National Laboratories. For the past four years she has been responsible for Sandia's program development in catalysis, which has included organizing a workshop on the future of catalysis research needs for Vision 2020: The U.S. Chemical Industry, a recent symposium on Department of Energy laboratory catalysis research investigating partnerships within and among the laboratories, and she has also visited several European catalysis organizations. She is active in the American Chemical Society (ACS) as chair elect of the Catalysis Secretariat, immediate past chair of the Industrial and Engineering Chemistry Division, and a consultant to the Committee on Science.

Dr. Jackson's research interests lie primarily in the area of structure and property relationships for heterogeneous catalytic materials, and she has investigated metal, metal oxide, and metal phosphate systems. Her work has included the study of a variety of catalytic processes including natural gas conversion to alternative fuels and commodity chemicals production. Her research often involves collaboration with universities and industrial firms.

Dr. Jackson has a B.S. degree in chemistry from George Washington University and a Ph.D. degree in chemical engineering from the University of Texas, Austin. Prior to attending graduate school she held positions at the ACS, as a U.S. Senate staff member, and the American Foreign Service Association at the U.S. State Department. She is active in issues regarding Native American education and is now chair of the ACS Committee on Minority Affairs. Jointly with her position at Sandia, she is presently an associate professor in the Department of Chemical and Nuclear Engineering at the University of New Mexico and serves on numerous advisory and editorial boards.

Henry T. (Hank) Kohlbrand is director of external technology and intellectual asset management for the Dow Chemical Company. He is one of eight directors responsible for corporate R&D at Dow. He assumed this position in 1997 and is responsible for Dow's external R&D programs, offices of science and technology in Europe and Japan, and the Intellectual Asset Management Technology Center. Previous responsibilities included corporate director of waste elimination, R&D director for engineering science and process development, and director of the Central Research Engineering Laboratory. Over his career he has worked in a number of research and manufacturing roles. In addition, he is recognized as an international expert in reactive chemicals engineering and process scale-up and has authored many publications in this area. He received a B.S. degree in chemical engineering from Illinois Institute of Technology in 1973.

Todd R. La Porte is professor of political science at the University of California, Berkeley (UCB) since 1965. Receiving his M.A. and Ph.D. degrees from Stanford University in 1962, he has taught public administration, organization theory, administrative behavior, and technology and politics at the University of Southern California and Stanford University as well as at UCB. He has published in the areas of organization theory, technology and politics, and the organizational and decision-making dynamics of large, complex, technologically intensive organizations, as well as public attitudes toward advanced technologies and the problems of governance in a technological society. He was a principle of the Berkeley High Reliability Organizations Project, a multidisciplinary team that has studied the organizational aspects of safety-critical systems such as nuclear power, air traffic control, and nuclear aircraft carriers. His current research concerns the evolution of large-scale organizations operating technologies demanding a very high level of reliable (nearly failure-free) performance and the relationship of large-scale technical systems to political legitimacy, especially in the nuclear domain.

He was elected to the National Academy of Public Administration in 1985, and he was a fellow at the Woodrow Wilson International Center for Scholars, Smithsonian Institution, and a research fellow at the Wissenschaftszentrum (Sciences Center) Berlin and the Max Planck Institute for Social Research, Cologne. He has been a member of the Board on Radioactive Waste Management and panels of the Committee on Human Factors and the Transportation Research Board of the National Academy of Sciences. He served on the Secretary of Energy Advisory Board, Department of Energy, and chaired its Task Force on Radioactive Waste Management, examining questions of institutional trustworthiness. He was a member of the National Research Council's panel on Human Factors in Air Traffic Control Automation and the Technical Review Committee for the Nuclear Materials Technology Division, Los Alamos National Laboratory.

David C. Mowery is the Milton W. Terrill Professor of Business Administration at the Haas School of Business at UCB. He has been at Haas since 1988, after work at Carnegie Mellon University, the National Research Council, and the Office of the U.S. Trade Representative. His professional interests are in technological change, international trade, U.S. technology policy, and the relationship between public policies and the private sector. He received a B.A. degree in 1974 and a Ph.D. degree in economics from Stanford University in 1981.

Lura J. Powell is the director of the Advanced Technology Program (ATP) at the National Institute of Standards and Technology (NIST). ATP fosters economic growth in the United States by accelerating the development of new technology. ATP is currently leveraging more than $1 billion in industrial R&D through projects proposed by industry, cost-shared with industry, and selected through a rigorous merit-based process.

A 23-year veteran of NIST laboratory research programs, Dr. Powell was chief of the Biotechnology Division at the NIST from 1991 to 1995 where she managed programs in DNA technologies, bioprocess engineering, biosensor technology, and structural biology. Dr. Powell served as chair of the Biotechnology Research Subcommittee of the National Science and Technology Council, which coordinates all federal biotechnology research, from 1993 to 1995. She has also served as chair of the Board of Overseers of the Center for Advanced Research in Biotechnology (a joint venture between the University of Maryland, NIST, and Montgomery County, Maryland) and as vice chair of the American Society for Testing and Materials Committee E-48 on Biotechnology. Her international activities included serving on the U.S. delegations to the U.S.–European Commission Task Force on Biotechnology and the Organization for Economic Cooperation and Development Working Party on Biotechnology.

Dr. Powell has a Ph.D. degree in chemistry from the University of Maryland. She is a member of the Federal Senior Executive Service and a member of the Board of Directors of the ACS. She is the recipient of the Department of Commerce Silver Medal Award and the Outstanding Public Service Award of the Montgomery County chapter of the International Personnel Management Association.

Christine S. Sloane is technical director for the Partnership for a New Generation of Vehicles and principal research scientist at General Motors Corporate Research and Development Center. In the former capacity, she is responsible for guiding and implementing the development of energy conversion and materials technologies in collaboration with the U.S. government and domestic automakers to enable the design and manufacture of highly energy-efficient vehicles. Her earlier research interests included manufacturing process emissions, air quality and visibility, and aerosol physics. Dr. Sloane has authored over 80 technical papers and co-edited one book. She received a Ph.D. degree from the Massachusetts Institute of Technology in chemical physics and is the recipient of General Motors' John C. Campbell Award for research. She is a member of the American Association for Aerosol Research, the American Association for the Advancement of Science, the Society for Automotive Engineers, Phi Beta Kappa, Sigma Xi, and formerly the editorial board of *Aerosol Science and Technology*.

John C. Tao is the director of Corporate Technology Partnerships at Air Products and Chemicals, Inc. He joined Air Products in 1974 in the research department of the company's Process Systems Group. He subsequently moved into the engineering area, where he assumed program responsibility for some of the company's early energy projects. In 1980 Dr. Tao became general manager of engineering for the company's International Coal Refining Company—a former Air Products joint venture. He subsequently was named director of planning and business development for Stearns-Catalytic World Corporation—a former engineering services subsidiary of Air Products—and in 1986 joined the Corporate

Science and Technology Center. Dr. Tao was named director of research and commercial development for the advanced separations department in 1989, and was appointed director of the department in 1991. He became corporate director of environmental, health, and safety assurance in 1992, and assumed his current position in December 1994. He received a B.S. degree in chemical engineering from the Carnegie Institute of Technology, an M.S. degree in chemical engineering from the University of Delaware, and a Ph.D. degree in chemical engineering from Carnegie Mellon University. He has authored nearly 40 papers and holds nine U.S. patents. Dr. Tao is a member of the Commercial Development Association, Licensing Executive Society, Industrial Research Institute, Council of Chemical Research, and the American Institute of Chemical Engineers. He is the current chair of Chemical Industry Environmental Technology Projects and a board member of the Penn State Research Foundation and the chair of the Management Committee of the Air Products–Imperial College Strategic Alliance and the Air Products–Penn State Research Alliance. He served on the advisory council for the Chemical Engineering Department of the University of Pennsylvania, and is active in the Lehigh Valley Business-Education Partnership.

Matthew V. Tirrell is the dean of the College of Engineering at the University of California at Santa Barbara. He was previously professor and head of the Department of Chemical Engineering and Materials Science at the University of Minnesota, where he served as director of its Biomedical Engineering Institute. He received a B.S. degree from Northwestern University and a Ph.D. degree from the University of Massachusetts. His interests are in transport and interfacial properties of polymers, with particular emphasis on molecular-scale mechanical measurements, bioadhesion, and new materials development. He is a member of the National Academy of Engineering.

William A. Wakeham has been the pro rector (research) at Imperial College since 1996, having previously spent eight years as head of the Department of Chemical Engineering and Chemical Technology. In addition, he became deputy rector of the college in 1997 and pro rector (resources) in 1999.

Professor Wakeham is interested in the relationship between the bulk thermophysical properties of fluids and the intermolecular forces between the molecules that comprise them. Thus, at one extreme, he is involved in the determination of intermolecular forces from measurements of macroscopic properties and the development and application of the statistical mechanics and kinetic theory that interrelate them. He is also actively involved in the measurement of the thermophysical properties of fluids under a very wide variety of thermodynamic states. The same thermophysical properties find application in the process industries within the design of a plant. A part of Professor Wakeham's activities are therefore concerned with the representation and extension of a body of accurate information on thermophysical properties in a fashion that allows their use with software packages for process simulation.

Professor Wakeham is the author or editor of 6 books and over 300 papers in the field of thermophysics; he is the editor of the *Journal of Chemical Thermodynamics* and the chair of Commission I.2 on Thermodynamics of the International Union of Pure and Applied Chemistry and of its Subcommittee on Transport Properties.

Appendix C

Origin of and Information on the Chemical Sciences Roundtable

In April 1994, the American Chemical Society (ACS) held an Interactive Presidential Colloquium entitled "Shaping the Future: The Chemical Research Environment in the Next Century."[1] The report from this colloquium identified several objectives, including the need to ensure communication on key issues among government, industry, and university representatives. The rapidly changing environment in the United States for science and technology has created a number of stresses on the chemical enterprise. The stresses are particularly important with regard to the chemical industry, which is a major segment of U.S. industry, makes a strong, positive contribution to the U.S. balance of trade, and provides major employment opportunities for a technical work force. A neutral and credible forum for communication among all segments of the enterprise could enhance the future well-being of chemical science and technology.

After the report was issued, a formal request for such a roundtable activity was transmitted to Dr. Bruce M. Alberts, chairman of the National Research Council (NRC), by the Federal Interagency Chemistry Representatives, an informal organization of representatives from the various federal agencies that support chemical research. As part of the NRC, the Board on Chemical Sciences and Technology (BCST) can provide an intellectual focus on issues and fundamentals of science and technology across the broad fields of chemistry and chemical engineering. In the winter of 1996, Dr. Alberts asked BCST to establish the Chemical Sciences Roundtable to provide a mechanism for initiating and maintaining the dialogue envisioned in the ACS report.

The mission of the Chemical Sciences Roundtable is to provide a science-oriented, apolitical forum to enhance understanding of the critical issues in chemical science and technology affecting the government, industrial, and academic sectors. To support this mission, the Chemical Sciences Roundtable will do the following:

---

[1] *Shaping the Future: The Chemical Research Environment in the Next Century,* American Chemical Society Report from the Interactive Presidential Colloquium, April 7-9, 1994, Washington, D.C.

• Identify topics of importance to the chemical science and technology community by holding periodic discussions and presentations, and gathering input from the broadest possible set of constituencies involved in chemical science and technology.

• Organize workshops and symposia and publish reports on topics important to the continuing health and advancement of chemical science and technology.

• Disseminate the information and knowledge gained in the workshops and reports to the chemical science and technology community through discussions with, presentations to, and engagement of other forums and organizations.

• Bring topics deserving further, in-depth study to the attention of the NRC's Board on Chemical Sciences and Technology. The roundtable itself will not attempt to resolve the issues and problems that it identifies—it will make no recommendations, nor provide any specific guidance. Rather, the goal of the roundtable is to ensure a full and meaningful discussion of the identified topics so that the participants in the workshops and the community as a whole can determine the best courses of action.